Juliana Dummer
Roberto Verdum

Soil Erosion

Juliana Dummer
Roberto Verdum

Soil Erosion

Analysis of Environmental Conditions and Agricultural Uses in the Occurrence of Linear Erosion in Chuvisca, Rio Grande do Sul/BR

ScienciaScripts

Cover image: www.ingimage.com

This book is a translation from the original published under ISBN 978-3-330-75778-3.

Publisher:
Sciencia Scripts
is a trademark of
Dodo Books Indian Ocean Ltd. and OmniScriptum S.R.L publishing group

120 High Road, East Finchley, London, N2 9ED, United Kingdom
Str. Armeneasca 28/1, office 1, Chisinau MD-2012, Republic of Moldova, Europe
Printed at: see last page
ISBN: 978-620-8-29233-1

SUMMARY

DEDICATION

I dedicate this work to God, my strength, my sustenance, without whom I would not have been able to overcome all the adversities and get this far, with whom I will follow wherever I am and as far as He enables me.

ACKNOWLEDGMENTS

To my advisor, Dr. Roberto Verdum, for his tireless and dedicated guidance in this research and for his friendship.

The institutions that made this dissertation possible: UFRGS and CAPES/Reuni for the two-year grant.

To Prof. Dr. Edinei Koester and Prof. Dr. Dirce Suertegaray for their suggestions for the development of this work.

The 1st Army Survey Division for donating the aerial photographs and topographic maps of the municipality of Chuvisca.

To my family, parents, brother, husband and godchildren who have supported me through this stage with their love and understanding, to whom I owe my greatest debt.

SUMMARY

Linear erosion is one of the forms of water erosion that causes land degradation, through furrows and gullies that reach the advanced stage of gullies. When there is no human intervention, they lead to the loss of arable land and Permanent Preservation Areas (PPAs), the interruption of roads, the contamination of watercourses and also become areas at risk of landslides, accidents and animal deaths. Furthermore, in agricultural areas, they generate excessive costs for municipalities and farmers, in terms of chemical nutrients and hours of work, with the use of machinery to maintain roads and crops. This research proposes an analysis of the conditioning factors of the environment (geology - lithology and structure, geomorphology, soil types and land use) in relation to the occurrence of linear erosion processes (gullies/gullies) in the municipality of Chuvisca, RS. The study is justified because there was a significant occurrence of gullies and gullies identified and registered between 2008 and 2010. The current results have provided greater detail on the genesis of linear erosion in the municipality. It is worth noting that of the 32 cases of this type of erosion that were mapped, all follow the direction of the set of lineaments and tectonic faults, in the main direction NE/SO and the secondary direction NO/SE. Thus, it can be said that linear erosion processes are partly conditioned by local-regional tectonics, both in their location on the ground and in their evolution. In addition, 66% of them occur in areas of conventional tillage, without soil conservation practices, and 28% in degraded Permanent Preservation Areas. In some cases, the areas affected by erosion have the same characteristics as their surroundings, leading us to believe that land use and inadequate agricultural practices are the main causes of erosion in the municipality of Chuvisca.

Keywords: Environmental constraints. agricultural uses and practices. linear erosion.

Municipality of Chuvisca.

1 INTRODUCTION

There are different forms of environmental degradation and, according to Guerra *et al.* (2007), these are related to the various components of a land unit: atmosphere, vegetation, soil, geology and hydrography. However, the degradation of soils becomes relevant, since soils are not easily replaced, as their formation and recovery processes are slow. According to the same author, the growth of food production and its incompatibility with environmental recovery has been one of the factors behind soil degradation, since it is linked to the degradation of the environment through the intensive use of pesticides and fertilizers, as well as agricultural machinery and even access routes to rural properties.

Despite being a vital resource, just like water, soil has been misused. According to the latest FAO (Food and Agriculture Organization of the United Nations) report in 2011, 25% of the world's land is degraded. The report presents the first global assessment of the world's land resources. A quarter of these resources are highly degraded. Another 8% is moderately degraded, 36% is in a stable or slightly degraded state and 10% is classified as "improving" land. The remaining surface of the planet is either bare (around 18%) or covered by inland bodies of water (around 2%).

One of the most significant forms of soil degradation is water erosion, which can cause economic, environmental and social damage. According to Bahia *et al.* (1992), Brazil loses around 600 million tons of soil every year due to erosion. Erosion not only causes soil impoverishment and a decline in agricultural productivity, but also increases production costs due to the need for a greater supply of nutrients for agricultural production. Erosion can cause social problems, because according to Bertoni and Lombardi Neto (2005), it gradually makes the land uninhabitable and, in more serious cases, it can cause the population to move away, since as soon as the soil is exhausted as a result of erosion, human societies tend to move to more productive land.

Human societies have also contributed to soil erosion by removing vegetation, overexploitation, overgrazing, agricultural and industrial activities, among others. In this way, social action has been one of the fundamental factors in accelerating erosion processes.

The natural or conditioning factors of the environment, such as the characteristics of the relief, the slope and the presence of lineaments and faults, the concentration of high rainfall rates and the physical and chemical characteristics of the soil increase its susceptibility to erosion. This susceptibility favors different forms of erosion, depending on the degree of loading of soil particles (EMBRAPA, 2006).

Among the different forms of water erosion, there are those caused by shallow laminar flow, where there is no deformation of the terrain, but rather gradual removal of the soil layers, and linear erosion characterized by concentrated flow, where there is deformation of the terrain in the form of furrows, gullies and gullies. Furrows are shallow channels originated by surface flow routes and gullies are deep surface channels, both of which have V-shaped channels. Gullies, however, are erosive features resulting from the evolution of gullies and furrows, and are caused by both concentrated surface and subsurface flow, and can reach hundreds of meters in length and width and tens of meters in depth. It differs from gullies in the presence of branches (side channels) and a U-shaped section. This type of erosion can have consequences for the population and the environment, such as the loss of usable area, silting up of watercourses and bodies of water and even the death of animals due to accidents.

Depending on the region of the country, gullies are also known as bossorocas and/or grotas, among others. The word boçoroca, according to the free dictionary of geosciences, derives from the Tupi, *iby-soroc*, (*iby* = *earth* and *soroc* = crack) meaning crack, rupture in the earth. According to Daee (1989), the formation of such features can occur due to a lack of planning and management of rainwater, such as the construction of roads, fences, infrastructures, with the torrent being directed to a single point and without an energy dissipation strategy, among others. However, according to Guerra *et al.* (2005), this is not the only process that causes gullies to form. Another existing erosion process is subsurface runoff, which forms concentrated flows in the form of tunnels or *pipelines* called *piping*, which can cause the surface above them to collapse, forming gullies in a short space of time.

Researchers from different countries are concerned about cases of gully erosion, which are increasing every year, especially in the light of climate change, which could aggravate the problem. Events such as the 2nd International Congress on Soil Erosion by Gullies on Global Climate Change, held in 2002 in the Chinese city of Chengchu, in the province of Sichuan, have highlighted this global problem, which has not been limited to rural areas. Data from the diagnosis carried out by Mendonça *et al.* (2001) shows the existence of 12 gullies in the urban area of the city of São Luiz, in Maranhão.

Other places such as the Cerrado of Santa Filomena in Piaui, the state of Minas Gerais and Mato Grosso also have problems with gully erosion, respectively studied by technicians from the Companhia de Desenvolvimento dos Vales do Sâo Francisco e Parnaiba (Codevasf), the Instituto Voçorocas and EMBRAPA Gado de Corte.

Ravines and gullies have also been classified by type. Sâo Paulo (1990) classified gullies as those caused by hydrological disturbances, larger and branched (subsurface runoff), and those produced by concentrated urban runoff (rainwater and wastewater) and rural runoff (road drainage and soil management) motivated by the land use factor. Pousen (1993) took into account where erosive processes occur, such as terrace erosion, and how long they last, such as ephemeral gullies. Other authors, such as Oliveira (1999), classify ravines and gullies as connected or disconnected from the drainage network. The systematization adopted by Oliveira (1999) for the study of linear erosive processes connected and disconnected to the drainage network considered the registration of erosive features in order to identify different erosive processes. In this sense, the presence of hierarchical subvertical fillets, spontaneous liquefaction, detachment of tracer cracks, erosion by infiltration or outcropping of the water table, *piping*, rain-splash erosion, diffuse and concentrated runoff inside and on the edges of the gully and mass movements are recognized in gullies and gullies connected to the drainage network. The presence of mass movements and surface runoff is emphasized for ravines and gullies disconnected from the drainage network. Finally, Oliveira considered as a second typology those ravines and gullies that present the integration of the various mechanisms mentioned.

In the municipality of Chuvisca (RS) there is a significant occurrence of ravines and gullies identified and registered between 2008 and 2010, in a study that made it possible to draw up a map of the spatial distribution of linear erosion and a detailed characterization of one of the cases identified. With a view to further detailing the genesis of these processes, this research proposes an analysis of the environmental factors (lithological/structural geology, geomorphology, soil types and land use) in relation to the occurrence of linear erosion processes (gullies/voçorocas) in the municipality of Chuvisca, RS.

1.1. Justification

Among the many forms of environmental degradation, soil erosion is the most common. Soil, like water, is a vital resource for humanity and is slow to form. Erosion leads to the loss of topsoil, reduces fertility and, in turn, primary production capacity, as well as water retention. Furthermore, in small towns like Chuvisca, which depend directly and indirectly on agricultural production, it is a potentially greater problem. In Chuvisca, soil erosion has caused significant degradation of the land, through furrows and gullies that have reached the advanced stage of gullies. Without any kind of intervention, they are causing the loss of arable areas and those defined as Permanent Preservation Areas (PPAs), the interruption of roads, the contamination of watercourses and even becoming areas at risk of landslides, accidents and animal deaths. In addition, they cause damage due to the excessive costs to the municipality and farmers, in terms of working hours, with the use of machinery to maintain roads and gullies.

It is also considered that this work, in addition to presenting an analysis based on the conditioning factors of the environment and its fragilities related to the occurrence of ravines and gullies in the Municipality of Chuvisca, will provide technical information on this environmental problem, serving to support the adoption of public policies and environmental education by the local government.

1.2. Study Area

The municipality of Chuvisca is part of the South Central Region of Rio Grande do Sul and is located at coordinates 30°45'17.56" south latitude and 51°58'7.56" west longitude. Chuvisca is part of the Camaqua micro-region and the Porto Alegre meso-region (map 1).

It has a population of 4,994 inhabitants, 323 living in the urban area and 4,671 in the rural area, with a population density of 22.42 inhabitants per km^2 , according to data from the 2010 IBGE census. The municipality's territory covers an area of 220 km2, representing 0.0815% of the state, 0.0389% of the region and 0.0026% of the entire Brazilian territory. Chuvisca's Human Development Index (HDI) was 0.776 in 2010, according to IBGE data[@] Cidades.

Chuvisca is a fundamentally agricultural municipality. Its economy is based almost exclusively on tobacco production, practiced on small rural properties, usually on plots of 2 to 6 hectares, employing an essentially family workforce. According to 2009 data from the IBGE[@] , 8.5 thousand tons of tobacco are produced annually.

MAPA DE LOCALIZAÇÃO DO MUNICÍPIO
DE CHUVISCA - RS

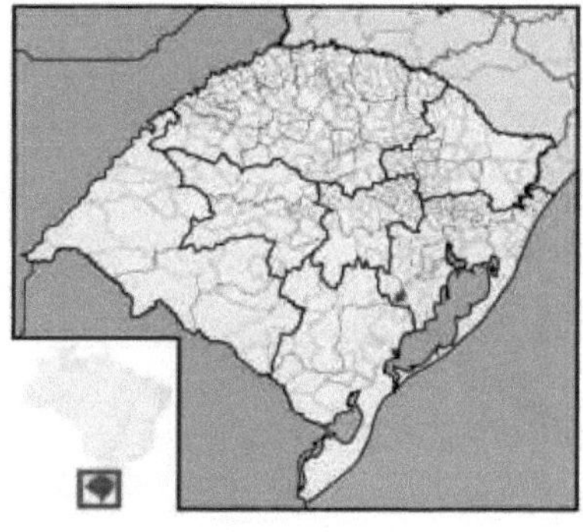

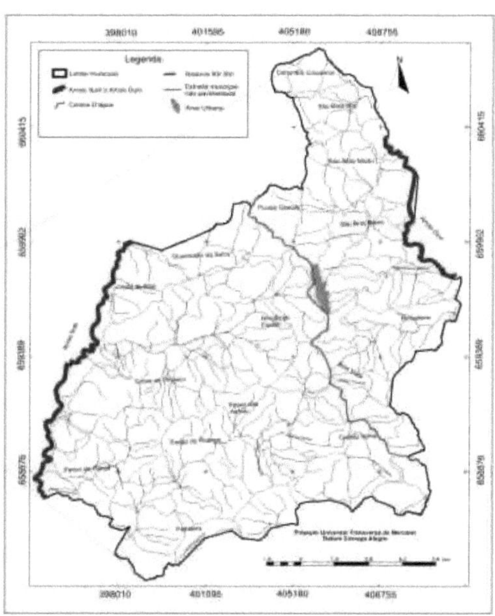

Map 1: Map of Brazil with the location of Rio Grande do Sul (in red); Map of Rio Grande do Sul with the delimitation of the municipalities, regional division and the location of the Municipality of Chuvisca (in red); and Map of the Municipality of Chuvisca with the road network, hydrography, and location of the urban area. Source: Prepared by the author, 2013.

2. OBJECTIVES

2.1. General

To analyze the geological and geomorphological conditions and soil vulnerabilities, as well as its occupation and agricultural uses, in order to understand the occurrence of erosion in the municipality of Chuvisca, RS.

2.2. Specifics

- Identify the relief units present in the study area as a result of past tectonics, the lithological context and exogenous processes;
- Identify soil types and their structural and textural characteristics as indicators of vulnerability to the occurrence of gullies and gullies;
- Conduct a survey of the main forms of occupation and current agricultural uses, considering these as triggers of erosion processes, as well as protectors in relation to them;
- Mapping and establishing a typology of water erosion processes in the municipality.

3. THEORETICAL BACKGROUND

3.1. Erosion phenomenon

Erosion is a natural phenomenon that acts in the processes of shaping the landscape in all of the Earth's ecosystems and is characterized by a process of removal, transport and deposition of sediments from the higher areas of the relief to the lower areas, such as plateaus and riverbanks. In the natural dynamics of the landscape, erosion can be caused by the force of water, characterizing pluvial, fluvial and/or glacial erosion, or by the force of the wind, characterizing aeolian erosion. However, this process of natural landform modeling, when intensified by society, has led to environmental degradation, loss of agricultural land and even social damage.

Water erosion, the focus of this study, is characterized by different mechanisms of action which, according to Bertoni and Lombardi Neto (2005), in the process of disaggregation, transport and deposition generate three main forms of water erosion: splashing, laminar (gullies) and linear (furrows, gullies and gullies).

Erosion caused by the impact of raindrops, popularly known as splash erosion, is the first form of water erosion. According to Cassol (2007), when raindrops hit the ground at a high speed, they cause the particles in the soil mass to break up. This process of disintegration and transportation causes the soil particles to be broken up and thrown in all directions by the splash effect. The erosive potential of disintegration is even greater as the average diameter of the raindrop increases, which is related to its kinetic energy and also to its intensity.

According to Cassol (2007), after the particles have broken down, the transportation process begins, which can take place by shallow laminar flow or concentrated flow in furrows.

As soon as the rainfall rate exceeds the infiltration rate of water into the soil, excess water will be deposited on the surface. If this excess is not intercepted, retained or stopped on the surface, suspended sediment will be transported. The same will happen through shallow laminar flow in the regions between gullies. In this case, we have what is known as gully erosion or laminar erosion, which causes the surface layers of the soil to be removed gradually. Because it is little noticed and neglected, it becomes the most damaging form of erosion. In areas where there is shallow laminar flow, there is also the process of disintegration, which predominates due to the impact of raindrops and transport by shallow laminar flow, thus characterizing an area of gully erosion (ELLISON, 1947; MEYER *et al.* 1975).

When moving water flows are concentrated in channels (furrows), often as a result of sloping terrain and/or small irregularities in the slope of the land, erosion by concentrated flow or linear erosion occurs.

According to Bertoni and Lombardi (2005) the upward movement of the water loosens the particles by lifting action. This detachment by abrasion occurs when the particles already in transit in the torrent strike or drag other particles on the surface of the ground, setting them in motion. The amount of material transported will depend on the transport capacity of the torrent, which in turn is influenced by the size, density and shape of the soil particles and the slowing effect of vegetation and other obstacles. These natural channels can reach larger proportions, depending on the volume and speed of the torrent. Where the surface flow is concentrated, both disaggregation and transport are carried out by the concentrated flow, characterizing an area of furrow erosion (ELLISON, 1947; MEYER E *et al.*, 1975). In their initial phase they can be undone with normal tillage operations, but in more serious stages they can prevent traffic or machinery from working on crops and roads.

Ravines and gullies are therefore the most critical stage of erosion by concentrated flows, when the furrows are widened by the displacement of large masses of soil, forming large cavities in length and depth. These cavities can be hundreds of meters long and tens of meters deep. Gullies are highly visible forms of water erosion that can affect soil productivity, restrict land use and even threaten roads, fences and buildings. Erosion on this scale can also cause siltation of watercourses, waterways, culverts, dams and reservoirs. In addition, suspended sediments, which have nutrients and/or pesticides associated with them, can damage and affect water quality. These fine materials, such as colloidal clay, remain in suspension and can also pollute underground aquifers, watercourses and affect aquatic life in general (CAREY, 2006).

According to Carey (2006), gullies are deep trenches dug into the ground, usually associated with places where there are concentrated flows of water in a furrow or channel, usually downhill, at speeds sufficient to remove and transport soil particles, gradually eroding the subsoil and potentially reaching the water table. It can also form where there is a natural concentration of runoff, such as at the headwaters of drains and embankments of slopes, or in places with relatively ephemeral water flows during heavy rainfall.

Gullies therefore develop in watercourses, roads, crops and/or other places where there is concentrated water run-off. In areas of cultivation or pasture, erosion of river and irrigation channels can develop into gullies if no protective measures are taken. Similarly, paths formed by livestock and roads can be a starting point for a small furrow, which can develop into a gully.

In addition, watercourses in a state of equilibrium with the right size, shape and gradients can lead to the emergence of erosion processes in the form of channels, which reach the advanced stage of gullies when they are disturbed by larger than normal flows. However, widening can occur by undermining the side walls of the gully. This collapse is facilitated when the walls are sloping and have saprolitic soils at the base.

In addition, the association of these factors with surface and subsurface runoff, which can percolate down the side ravines in the form of ducts, contributes to the formation of secondary gullies or ramifications, thus increasing the proportions of erosion. Another point where gullies develop is in places where there is constant dripping, such as springs or waterholes or drainage lines, which saturate the soil, making its structure weak and susceptible to erosion (CAREY, 2006). This constant humidity can also weaken the vegetation and thus offer less resistance to erosion and contribute to its advance.

The importance of studying the phenomena associated with the formation of ravines and gullies lies in establishing prevention and control measures, as well as techniques compatible with combating the problem, which is usually difficult and costly. They are also justified on the grounds of preserving soil and water quality, especially where there is a reasonable chance of success in controlling them and recovering the areas affected by them. Whether they are gullies or gullies on access roads, in the countryside, on crops or in permanent preservation areas (APP). It should also be noted that the prevention of such erosion is quicker and less costly than remediation.

3.2. The implications of the biophysical environment for erosion processes - natural and social factors

According to Ramalho (2003), the biophysical environment is the result of a combination of physical, chemical and biological elements that interact with each other in a given area. The study of the characteristics and dynamics of the environment is therefore considered in order to assess the interrelationships of human activities that appear to indicate imbalances. For this author, knowing the characteristics of precipitation, soils, slope morphology, among others, and the actions of society is fundamental to diagnosing and assessing the potential and/or fragility of the environment, given the implications of human activities.

Climate directly influences weathering rates, defining soil types according to topography and rocks. For some researchers, such as Thorner (1980) cited by Ramalho (2003), climate is the most important variable because it determines natural vegetation and contributes to higher or lower rates of soil erosion. Regular rainfall favors lush vegetation cover, which can lead to soils with lower densities and high water infiltration rates (drainage), which means less runoff and, consequently, lower erosion rates. According to Cassol (2007), among climatic factors, the most important in the process of water erosion is rainfall, since rainwater and surface runoff are the agents that cause disintegration and transportation. Bertoni and Lombardi (2005) also point out that rainfall intensity is the most important factor. The greater the intensity, the greater the loss through erosion when the infiltration rate of the soil is exceeded. Infiltration, which is the movement of water within the soil surface by the forces of gravity and capillarity, depends fundamentally on the force of gravity when the soil is saturated. However, in some cases the unavailability of precise local data makes it impossible to evaluate the climate variable in soil erosion studies.

According to Ramalho (2003), geological conditions not only provide the material that characterizes the configuration of the relief, but also determine the mineralogical characteristics of soils. Rocks, which can be magmatic, sedimentary or metamorphic, respond directly to the action of weathering agents, depending on their greater or lesser lithological resistance, determined by the intrinsic characteristics of each one.

Studies by Beavis (2000) on morphostructural geology show a significant correlation between the orientation of fractures and faults and the occurrence of gullies and gullies. This would indicate that there is a strong influence of the structures present in the rock substrate on both the development and orientation of gullies and gullies. Studies carried out, for example, in the northwestern Himalayas by Malik and Mohanty (2007), have shown a strong control of tectonic activity on the evolution of drainage and the landscape. This shows, according to Tricart (1997), the importance of knowing the characteristics of the elements that make up this relationship in order to explain the dynamic processes that shape the landscape.

Also, the study of landforms that influence the flow of water in different paths over the terrain has proved fundamental for characterizing and analyzing erosion processes. Through the geomorphology given by the shape of the profile and the plane, the degree of slope and the length of the ramp, as well as the relationship between these two in determining the topographic factor (LS), we have a set of important data for understanding soil erosion.

Authors such as Resende *et al.* (1997) have proven the importance of studying the shape of slopes and state that soil erosion increases from concave to convex pedoforms, passing through the linear one which has greater stability. Similarly, studies such as Wang *et al.* (2002) and Sanchez *et al.* (2009) have shown that in places where the soil and management practices are the same for the whole area, the shape of the slope has been shown to be responsible for the greatest soil losses, erosion risk and natural erosion potential.

The determination and study of relief compartmentalization, according to Casseti (2005), is an important aid to understanding geomorphological vulnerability and potential. Vulnerability, from a geomorphological perspective, means the erosive susceptibility of the relief, both under natural conditions and predicted as a result of certain uses or occupations, with the topographic compartment as a support or resource. Potentiality, as the name implies, refers to certain individualities that can be rationally appropriated for specific purposes, such as the allocation of areas containing deposits of cover with natural fertility to agricultural activities, or even special morphologies, such as canyons and faults, aimed at tourist exploitation. In turn, combining studies on the different degrees of vulnerability of the relief with its potential makes it possible to produce maps with indications for sustainable uses or for environmental protection, which would help municipalities to control soil erosion.

Soil is also a determining variable, since erosion is not the same in all types of soil. Physical properties such as structure, texture, permeability and density, as well as chemical, mineralogical and biological characteristics, according to Bertoni and Lombardi (2005) have different influences on erosion. Generally speaking, soil structure and particle arrangement should also be taken into account when studying erosion. Determining the physical and chemical properties of the clay, which make the aggregates stable in the presence of water, and the organic matter content, as well as retaining two to three times its weight in water, increases the infiltration of water into the soil and reduces losses through surface runoff.

Soil has several intrinsic properties that respond differently to the process of water erosion. This means that each type of soil has a certain degree of susceptibility to erosion. This susceptibility of the soil to erosion is called erodibility. According to Cassol (2007), the soil properties that influence erodibility by water can be separated into two groups: (a) those that resist the forces of dispersion, disintegration, abrasion and transportation of rain and surface runoff; (b) those that affect water infiltration capacity, permeability and total capillarity of water storage in the soil.

According to Cassol (2007), the first group includes the stability of aggregates, which is the most important property in determining the degree of disintegration when the soil receives the impact of raindrops. This stability is mainly related to the clay and organic matter content of the soil. According to Bertoni and Lombardi (2005), the dominant mineralogy also plays a role. Aggregates from soils with montimorillonitic clay are not very stable in water, while soils with kaolinitic clay are more stable, with illite in an intermediate position.

The size, shape and density of the particles are also decisive in the process of water erosion, as they affect their transport capacity. Small particles with a spherical shape are more easily transported than larger particles with an irregular shape. However, Cassol (2007) states that organic matter provides greater stability to aggregates, but lower particle density. When the soil disintegrates, the particles containing organic matter are more easily transported.

The properties that affect the second group, water infiltration, i.e. permeability and total water storage capacity, are mainly linked to soil structure. According to Cassol (2007), factors such as compacted layers, impermeable subsoil and soil depth are also important. According to Cassol (2007), well-structured soils have a higher infiltration capacity than unstructured soils.

According to Bertoni and Lombardi Neto (2005), the texture or size of the soil particles are determining factors in the greater or lesser amount of soil washed away by erosion. Sandy soils generally have more pore spaces of the macro-pore type (>0.05 mm), favoring the infiltration of rainwater without major damage. However, they may have a weak structure, mainly due to the lower proportion of clay, which would confer greater reactivity between the solid particles, stabilizing aggregates. As a result, soil loss due to erosion may be greater in this type of soil. Clay soils have a higher proportion of small pores (<0.05 mm), making it difficult for water to infiltrate and favoring surface runoff. However, they have a higher particle cohesive force, increasing their resistance to erosion. Therefore, these attributes can result in soils with lower erodibility.

The use and cover of the land determined by human occupation, through deforestation, agricultural cultivation, road construction, the creation and/or expansion of urban areas, among others, become important in triggering erosion processes, since they are closely related to waterproofing indices and the infiltration capacity of soils. Accelerated occupation, the subdivision of farms and ranches, deforestation and the destruction of hillsides to remove material for construction are also examples of human activities that can trigger erosion.

Studies presented by Bertoni and Lombardi (2005) have shown that even in rural areas, where the rate of soil sealing is insignificant, there is a considerable difference in soil loss due to erosion. This depends on the type of crop, with clear advantages where soil management is adopted with strip-till systems. Among the plants that stand out with the greatest amount of soil washed away are castor beans, cassava and beans, with peanuts, rainfed rice and cotton appearing in a second group. This is in line with studies by Boiffin and Bresson (1987) in north-west France, where it was shown that the appearance and development of gullies is not related to rainfall characteristics, but rather to the state of the soil in the fields, as a result of the ways in which it is managed and cultivated. Soil management and cultivation can therefore be considered to protect or induce erosion processes.

Research carried out by the Soil Conservation Section of the Agronomic Institute of Campinas on three types of soil in the state of São Paulo showed a difference in soil loss in different uses. In forest areas, soil loss is around 0.004 t/ha, while in coffee and cotton areas it is 0.9 and 26.6 t/ha, respectively (BERTONI and LOMBARDI, 2005). For Cassol (2007), in relation to land use for agricultural purposes and annual crops, the critical periods in relation to water erosion occur when the most intense rainfall coincides with periods when the soil is bare. In other words, when in the conventional tillage system, there is rain between tillage and sowing, or when the vegetation cover offers little protection. In this case, erosion losses are higher and the characteristics of the rain can aggravate the process.

Vegetation cover is the land's natural defense against erosion. According to Cassol (2007), it has an effect on the interception of raindrops, dissipating the kinetic energy of their fall, i.e. it serves as direct protection against the impact of raindrops. Vegetation also acts as a disperser of water, intercepting it and even allowing water to evaporate before it reaches the soil. This also contributes to the decomposition of plant roots, forming channels or biopores in the soil, increasing water infiltration when the soil is saturated. This improves the soil's structure through the addition of organic matter and increases its water retention capacity, reducing the speed of runoff by increasing friction on the surface. However, according to Cassol (2007) the extent of the protection provided by the vegetation cover depends on the type of plant and roots. Prostrate plants protect the soil better than erect plants, and broad leaves protect it better than narrow leaves, as long as the number of leaves is the same. The same author also points out that a compact and abundant root system, like that of most grasses, protects the soil better, promoting better formation of stable aggregates. By producing organic matter, vegetation, not only the roots but also the humus produced, helps to form more stable aggregates that are resistant to soil erosion.

4. MATERIALS AND METHODS

This study required three stages: (1) collection and analysis of existing data (2) field research, (3) laboratory and office work.

The stage of **collecting and analyzing existing data** concerns the construction of the theoretical-methodological framework for the study, based on consultation of books, theses, dissertations and articles on the subject of soil erosion, the phenomenon of gully and gully erosion and the variables that control these processes. Some references can be cited, such as: Beavis (2000); Viero (2004); Bertoni and Lombardi Neto (2005); Denardin *et al.* (2005); Valentin *et al.* (2005), Guerra *et al.* (2005) and (2007); Cassol and Lima (2007); among others. Materials were also consulted on methods for integrating and generating data in a geographic information system (GIS) in order to cross-reference geology, soil, geomorphology and land use data with data on the location of gullies and gullies already registered in the municipality of Chuvisca.

The map showing the location of gullies and gullies was used to analyze the erosion processes. This map, drawn up following information provided by schoolchildren and the local population, enabled a pre-registration form to be drawn up. With this information, several field campaigns were then carried out to identify the erosion processes. In each case, the UTM coordinates were noted and photographic records were taken. The field data (points and traces) were downloaded into the *GPSTrackMakerPro* program, version. 4.6, from which they were exported in *.kml and *.shp formats. The kml files were viewed in *Google Earth®* to improve the contours of the gullies. After this stage, the original points and the improved tracings, in shp format, were imported into *SPRING®* and a map of the location of 32 registered gullies was drawn up.

The maps of hydrography, lithological and structural geology, slope, hypsometry, relief units, soil types and land use were used to analyze the environmental constraints and erosion processes recorded.

The map of Chuvisca's hydrographic network was created by importing the hydrographic data from the 1:50,000 scale topographic maps, published by the Ministry of the Army, Directorate of Geographic Services and digitized, as mentioned above, available in Hasenack and Weber (2010). After importing the data, vector edits were made based on the CBERS image mosaic to improve the contours and current location of the watercourses.

The geology map of the Municipality of Chuvisca was based on the mapping carried out by CPRM - Geological Service of Brazil (CPRM, 2003 and 2007) and field surveys, through which a *SPRING®* file in *shape* format (shp) was produced for the delimitation of the local geology that gave rise to the 1: 50,000 scale map of the municipality.

The soil map was based on the IBGE Natural Resources Survey (1986) carried out by the RADAMBRASIL project on a scale of 1: 1,000,000. After importing the *shape* file (shp) into *SPRING®*, the process of editing and updating the identified soil classes to the new nomenclature began.

Using the Numerical Terrain Model (MNT) derived from SRTM (*Shuttle Radar Topographic Mission*) data, we created the hypsometric and clinographic maps of the Municipality of Chuvisca and obtained the data corresponding to the slope, vertical and horizontal curvature of the slope and the LS factor for each gully mapped in the municipality, both based on the method proposed by Camara *et al.* (1996).

The hypsometric or altitude map was divided into six classes represented in color: 40-80m, 80-120m, 120-160m, 160-200m, 200240, 240-280 and greater than 280m.

The clinical or slope map was drawn up in percentages, and the intervals adopted are based on the method used by EMBRAPA (SANTOS *et al.*, 2006): 0 to 3% flat, 3 to 8% gently undulating, 8 to 13% moderately undulating, 13 to 20% undulating, 20 to 45% strongly undulating, and over 45% mountainous. The slope of each gully was obtained from the map and the import of the points of the mapped gullies, which generated a table with the slope for each point of occurrence.

The vertical curvature derived from the MNT generated expresses the shape of the slope when observed in profile. According to Câmara *et al.* (1996), it is defined as the second derivative of altitude, which can be described as the variation in slope over a given distance. Translating these definitions into common perception, it refers to the convex/concave nature of the terrain when analyzed in profile. It is expressed as a difference in angle divided by horizontal distance, which can take different units.

According to Câmara *et al.* (1996), horizontal curvature expresses the shape of the slope when observed in horizontal projection. It is also defined as a second-order derivative, but not of the elevation, but of the contour lines. In analogy to the relationship between vertical curvature and slope, horizontal curvature can be described as the variation in the orientation of slopes over a given distance.

The topographic factor (LS), ramp length and slope grade were obtained using the equation proposed by Wischmeier and Smith (1978):

$$LS = [0.065 + 0.0456 \text{ X } S + 0.006541 \text{ X } S^2] \text{ X } (L \div 22.13) \text{ m}$$

where m = 0.2 for S < 1%; m = 0.3 for 1% ≤ S ≤ 3%; m = 0.4 for 3% < S <5%; m = 5 for S ≥ 5%, S - slope gradient (%) and Le - slope length (m). The S and Le parameters used in the Wischmeier and Smith (1978) equation were determined by the Williams and Berndt (1976) equation:

S=0.25.Z.(LC 25+LC50+ LC75) A

where Z: is the gradient between the control section and the highest point in the basin (m); LC25, LC50 and LC75: the lengths of the contour lines at 25, 50 and 75% of Z and A: the area of the basin (m).[2]

Le = LC LB

2 EP (LC2 LB2) 0,5

where LC: is the sum of the lengths of the contour lines of the hydrographic basin (m); LB: the sum of the lengths of the base contour lines (m) and EP: the number of extreme points (those that occur when a main channel or talvegue cuts the contour lines of the basin).

The land use and cover map was drawn up based on LANDSAT satellite images. After downloading the image in *tiff* format, it was transformed into *GRIB* in the Impima® program, to be opened later in SIG SPRING®. After georeferencing the image, contrast was applied to improve visual analysis of the image. Based on this contrast, the image was color-composed and saved as a synthetic image (B-Band 3, G-Band 5 and R-Band 4), making it possible to classify land use in the municipality. In order to map vegetation cover and land use, we used the techniques of visual interpretation of orbital images, considering the spectral responses of each band, and field knowledge was also important. Thus, the image was classified using the supervised region type, using the Bhattacharya classifier, whose acceptance threshold was 90%. After classification, class mapping was carried out, creating a thematic information plan for which it was possible to quantify land use in 5 classes. These are: native forest, afforestation, cultivation, exposed soil and field.

The **field stage** consisted of new field campaigns to observe the registered erosion zones, visits and interviews with the owners of these areas, in order to understand the evolution of these erosion processes and practices related to the conservation and recovery of degraded areas. Finally, this stage included successive field observations of the erosive mechanisms that control the evolution of the erosive features, for the selection of erosions, for the production of record sheets (annexes I, II, III, IV, V and VI) and the preparation of sketches illustrating the dynamics and shape of these features. This selection was based on the choice of erosion process typologies. In the case of the Municipality of Chuvisca, three typologies were identified: (i) gullies and gullies developed in areas of fields used for livestock farming; (ii) gullies and gullies developed in crop areas; and (iii) gullies and gullies developed on roads. It is necessary, however, to clarify that there is some divergence in the distinction between furrows, gullies and gullies. Technicians from the São Paulo Institute of Technology (ITP) have linked the definition of a gully to channels carved by the upwelling of the water table. This definition, however, does not take into account the process of surface water erosion and the different stages of erosion, usually related to furrows that evolve into gullies which, in turn, evolve into gullies. Authors such as Guerra *et al.* (2005), opt for the dimensional definition, widely used in the academic community, which distinguishes gullies as erosion incisions with a width and depth of more than 50 centimeters.

This study was based on the mapping carried out by Dummer (2011) who, for practical reasons, only recorded the most significant cases identified in the municipality. In other words, erosion incisions larger than 5 meters wide, 3 meters deep and 10 meters long, without checking whether they were gullies or gullies. In the current phase of the study, despite opting for a definition linked to the upwelling of the water table, most of the erosion processes have not been classified due to the impossibility of accessing the inside of them to verify the upwelling of the water table. This explains why some cases are referred to as gullies, others as gullies or linear erosion.

The **laboratory and office stage** involved analyzing the thematic maps: hydrological, geological (structural and lithological), soil taxonomic, hypsometric or altitude, clinographic or slope, topographic factor (LS), relief unit, land use and land cover, and then comparing them with the data on the location of the gullies and gullies identified in the municipality of Chuvisca. This stage also included cross-referencing this data in a GIS environment using the *Sprig* ® program. The comparison of this data aims to identify the possible controlling variables of the erosion processes found in the municipality. This stage also included the preparation of a structural geological map of the municipality based on the geological (lithological) map of the Municipality of Chuvisca (DUMMER, 2010), scale 1: 50.000 scale and structural data from the Geological Mapping of RS - CPRM (2007), at a scale of 1:750,000, as well as the preparation of the Relief Units map at a scale of 1: 50,000 based on the hypsometric and altitude maps and field observations.

The laboratory and office stage also included the analysis of aerial photographs of the study area. Photographs numbered 18576, 18575, 18867 and 16866 from 1964 on a scale of 1:60,000, made available by the Army's 1st Survey Division for stereoscopic study, were analyzed, as were satellite images from *Google Earth*®. The analysis of aerial photographs and satellite images based on field studies and in comparison with the structural geological map aimed to investigate the morphological scenario in which the municipality is located, as well as the probable existence of geological faults that could be the main triggering factor for the erosion processes identified.

Finally, the laboratory stage involved interpreting the field and laboratory data and writing a dissertation on the subject.

5. ENVIRONMENTAL FACTORS AFFECTING THE OCCURRENCE OF LINEAR EROSION IN THE MUNICIPALITY OF CHUVISCA

The aim here is to analyze the environment by investigating different aspects of the dynamics of nature in order to understand the mechanisms at work in the erosion processes in the municipality of Chuvisca.

According to Verdum (2013), the marks of production processes related to agricultural practices and the actions caused by natural phenomena generate morphogenetic processes that can be distinguished in their dynamics, as well as in their interactions and relationships with the fragilities of the environment. The study of these processes is relevant due to their complexity and the urgent need to rethink the ways in which various human activities are inserted, as well as the growing need for methodological and technical improvements in order to assess the potential and constraints of the environment in relation to these activities.

In the municipality of Chuvisca, a mapping of gullies and gullies (DUMMER, 2011) was carried out between 2010 and 2011 (map 2), with the aim of registering and quantifying existing erosive features, in order to study the conditioning factors of the environment. Of the 32 cases mapped, around 90% are at least 15 m wide and 25 m long. In some locations, such as Periquiteira, Sâo Bràs Mèdio and Baixo and Picada Grande, these erosion processes were not fully covered by the survey, so there may be other cases not reported in this work.

In order to study the conditioning factors of the environment, this mapping was then compared with the available regional data on geology (lithology and morphostructure), geomorphology (altimetry, slope, slope length, topographic factor and relief unit), soil types and, subsequently, land use together with field observations.

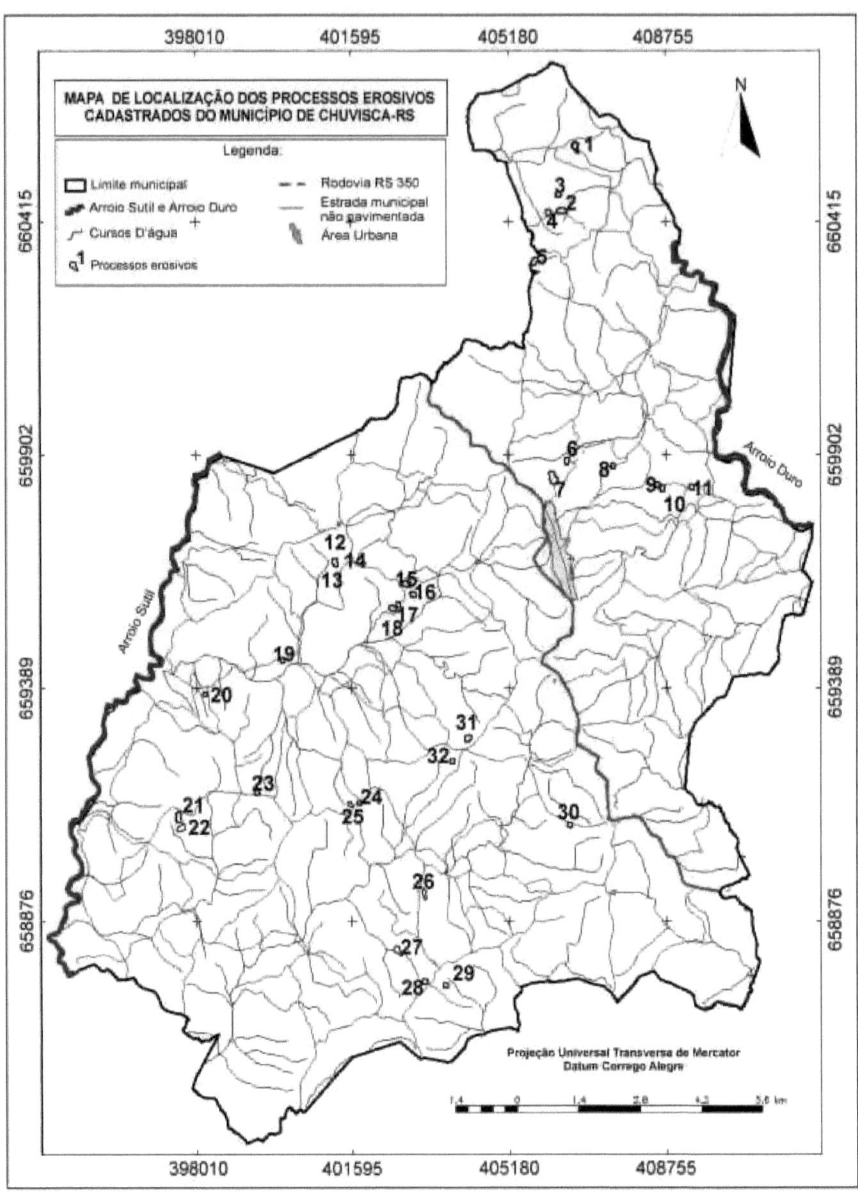

Map 2: Location map of the 32 linear erosion processes registered in the municipality of Chuvisca, RS. Source: Prepared by the author.

3.3. The regional tectonic context as an indicator of erosion processes in the municipality of Chuvisca

Chuvisca is located in the Sul-rio-grandense Shield in the region of the Pelotas Batolith, which, according to Philipp *et al.* (2000), is made up of igneous rock associations that mark the evolution of a plate convergence zone during the Neoproterozoic, and which are subordinately interspersed with medium and high-grade metamorphic terrains that make up the Paleoproterozoic basement areas. The boundary between these units is defined by direct shear zones on a continental scale, which were responsible for the segmentation of the Precambrian units and for their configuration as elongated bands in a NE-SW direction.

It was observed that the pattern of the main drainage network in the municipality is of the parallel type characteristic of an area under the influence of geological faults. The orientation of the lineaments observed in rock outcrops showed faults in the main direction NE-SW, and secondary SE/NW (fig.1).

Figure 1: Determining the orientation of faults (directions indicated by the clasts) in an outcrop of granitic rocks (Dom Feliciano Suite) in the locality of Sâo Bras Alto. Main: N76°E Secondary: N40°W. Photo: author's collection, 2014.

These typical structures of the region reveal the strong influence of the geological structure on the drainage network and, consequently, on the sculpting of the relief.

With regard to the occurrence of erosion processes, geological structures, according to

Beavis (2000), generally determine the location of erosion in the relief. In Chuvisca, from the analysis of aerial photographs and satellite images (fig. 2), based on the field study and the structural geological map (map 3), it was possible to identify that most of the erosion processes develop in the preferential directions associated with the set of main geological faults, i.e. NE/SW.

Figure 2: Cut-outs from aerial photograph 16867 (left) and satellite image covering part of the Municipality of Chuvisca with lineaments indicated (dotted red line), according to CPRM mapping (2007). Source: adapted from the 1st Army Survey Division and *Google Earth®*.

It was also found that these structures condition the evolution of erosion processes, since there are often faults in the rocks, with *slickensides*, on the inside walls of gullies and gullies, and it is at the limits of these faults that the slopes collapse in most of the erosion processes mapped (fig. 3). These planes of weakness are evidence of the past movement of the blocks, which are quite clayey and provide preferential zones for the sliding and collapse of the walls, contributing to the advance of erosion.

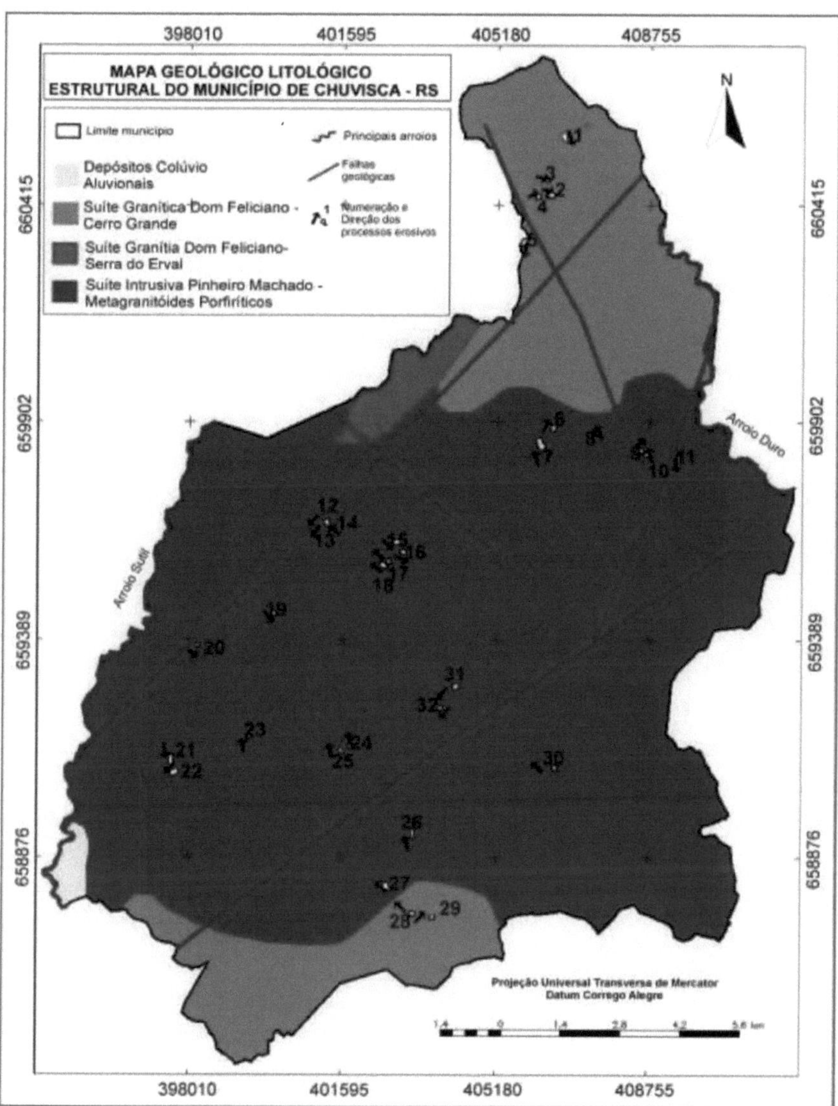

Map 3: Geological-Lithological-Structural Map of the Municipality of Chuvisca showing the predominance of the Pinheiro Machado Intrusive Suite, subordinately to the Dom Feliciano Granitic Suite and the presence of major faults in a NE-SW direction. Source: Adapted from the Geological Map of RS - CPRM (2007), scale 1:750.000.

Other studies, such as Valentin *et al.* (2005), have shown the same when they found that the development of relief can often be related to tectonically induced compressive or tensile forces, which in turn generate weakened areas that become starting points for the processes of weathering and the formation of subsurface ducts, which result in the subsidence of slopes.

It was also found that the lineaments are responsible for the orientation of the erosion processes observed in the field. As can be seen in figure 4, they develop preferentially in a NE/SO or NO/SE direction.

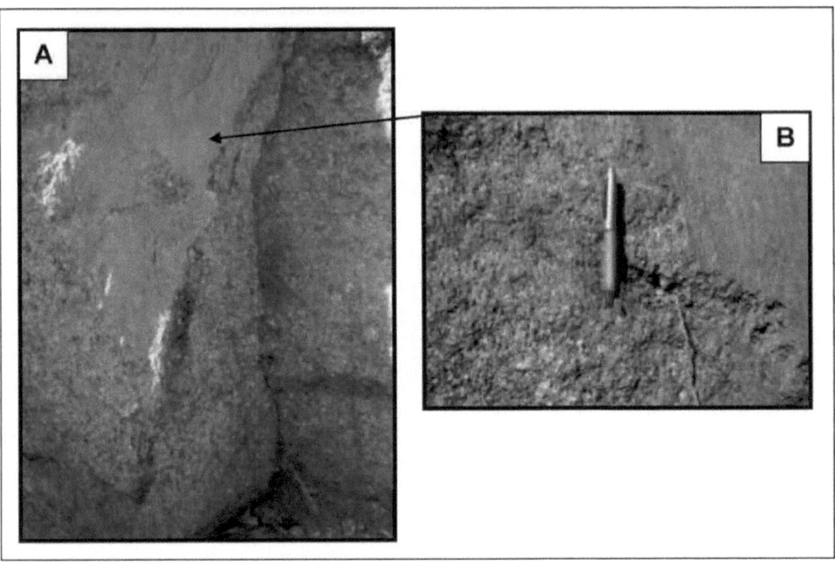

Figure 3: A and B (detail), landslide zones due to the presence of *slickensides*, generated by planes of weakness in the rock faults in the walls of the gully located in Sâo Bras Alto, Chuvisca - RS. Photo: author's collection, 2010.

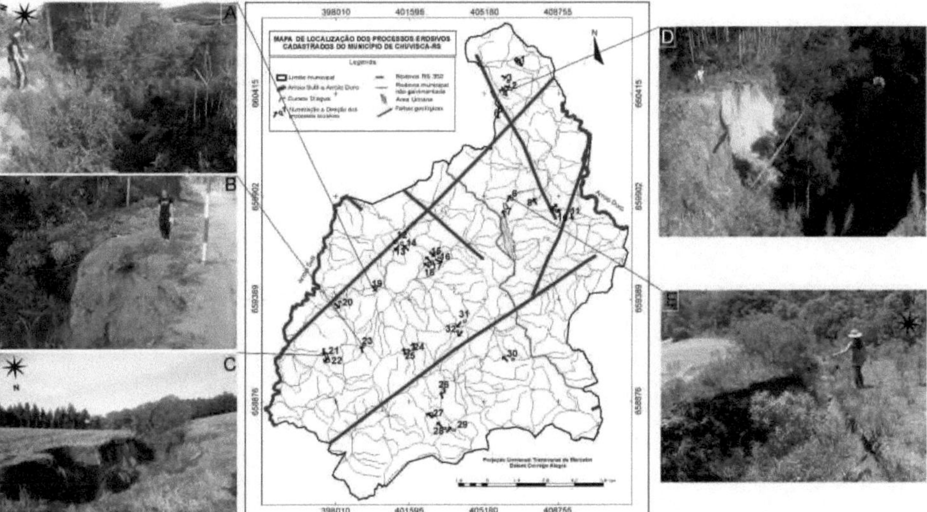

Figure 4: Examples of ravines and gullies registered in the municipality, developing in the direction of the lineaments: Photo A: NO-SE; B: NO-SE; C: NE-SO; D: NO-SE; E: NE-SO, as indicated by the location rocks on the map.

With regard to the regional lithological formation in Chuvisca, in addition to the predominant occurrence of granites as source material, metamorphic rocks occur subordinately. The main units in the study area are classified as: i) Xenoliths of metamorphic rocks (basement gneisses) ii) Pinheiro Machado Intrusive Suite, iii) Dom Feliciano Granitic Suite (Serra do Herval facies) and iv) Dom Feliciano Granitic Suite (Cerro Grande facies) (DUMMER *et al* 2010) (fig. 5 and 6).

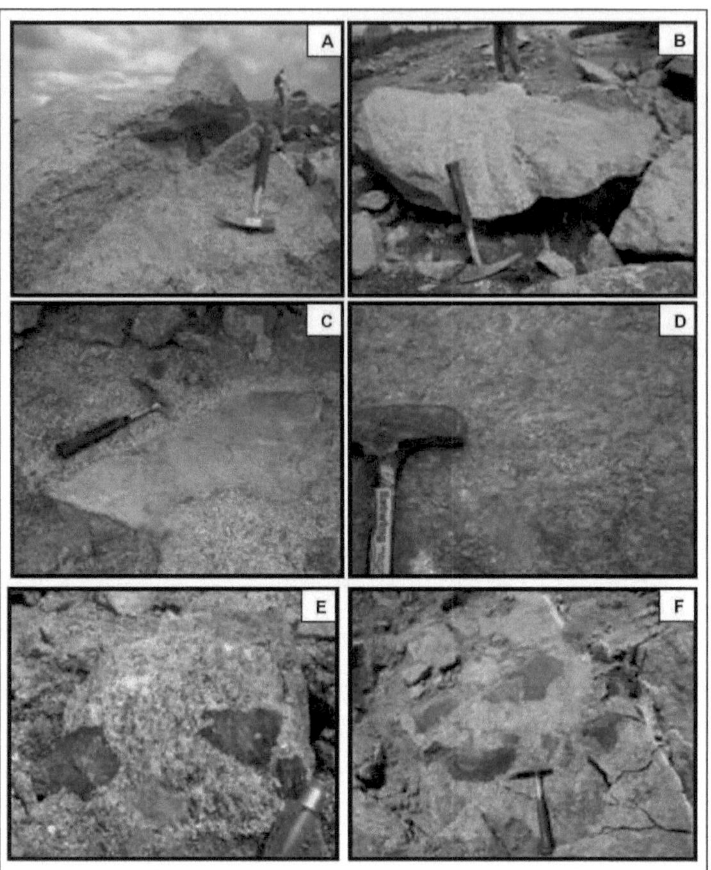

Figure 5: Points sampled in the geological survey in the municipality of Chuvisca, RS. A) Cerro Grande SGDF - pink granite; B) Light gray xenoliths ranging from 50 to 150 cm; C) Fragments of typical SIMP rocks; D) Cerro Grande SGDF - grayish-pink equigranular granites; E) Rare dark gray granitoid xenoliths smaller than 10 cm; F) Dark gray biotite-based xenoliths measuring approximately 50 cm;

Figure 6: Points sampled during the geological survey in the municipality of Chuvisca, RS. G and G1 (detail) SGDF- Cerro Grande, 20 to 40 cm xenoliths; H and H1 (detail) SIPM - whitish gray granodiorites, I) SIPM- pink granodiorites. Photos: author's collection, 2010.

Cross-referencing data from geological mapping and the occurrence of gullies and gullies in the Municipality of Chuvisca showed that 73% of the erosion features recorded occur in the Pinheiro Machado Intrusive Suite and 27% in the Dom Feliciano Granitic Suite (Cerro Grande facies). In the areas belonging to the Pinheiro Machado Intrusive Suite (SIPM), according to the CPRM classification (2007), whitish gray granodiorites, porphyritic gray granodiorites and veins of pink granodiorites measuring 30 cm thick are found. In the Dom Feliciano Granitic Suite, grayish-pink equigranular granites with lots of mica and a medium equigranular texture are identified (DUMMER *et al.* 2010).

According to Beavis (2000), the role of lithology in the development of water erosion processes also depends on mineralogy and particle size, which determine the intensity of erosion. According to Press *et al.* (2006), mineralogical composition is a major factor, since rocks can undergo relatively different rates of weathering depending on their mineralogical composition. Thus, over geological time, the action of exogenous processes in transforming rocks into sediments and/or soils, whether or not influenced by human action, can produce forms of soil degradation such as gullies and gullies.

Inferences made in the field show that, in many cases in the municipality of Chuvisca, erosion features develop on a highly weathered rocky substrate with a thick layer of saprolitic material. Studies on the relationship between geological formation and the spatial distribution and orientation of erosion incisions carried out by Silva *et al.* (2003), showed that once the saprolite is reached, there is a reassembling propagation guided by the geological structures (faulting and/or fracturing zones) of linear erosion.

In addition, the exposure of a thick layer of saprolite, characteristic of the erosion processes mapped in the Municipality of Chuvisca, makes it more vulnerable to erosion, since saprolite has low levels of organic matter and poor aggregate stability (DUMMER, 2011). As a result, it does not support the underlying soil layers, causing the slopes to collapse. Similarly, the lack of nutrients does not favor the development of plants that would help to contain erosion.

3.4. Geomorphological context as an indicator of erosion processes in the municipality of Chuvisca

The municipality of Chuvisca is located in the geomorphological province known as the Sul-rio-grandense Shield. The Sul-rio-grandense Shield represents the bedrock of the state of Rio Grande do Sul and is located in the central-southern region, occupying around 65,000 km^2 of area in the state (Chemale Jr., 2000) and is delimited to the north, west and southwest by the Paranà Basin, and to the east by the Pelotas Basin, also known as the Coastal Province of Rio Grande do Sul.

Based on regional data and field analysis, it was possible to determine that the Municipality of Chuvisca has three geomorphological compartments: ridges, hills and alluvial plains, with altimetric ranges varying from 40 to 349 m in elevation. The topographic profile (fig. 7) shows the different morphological features mapped.

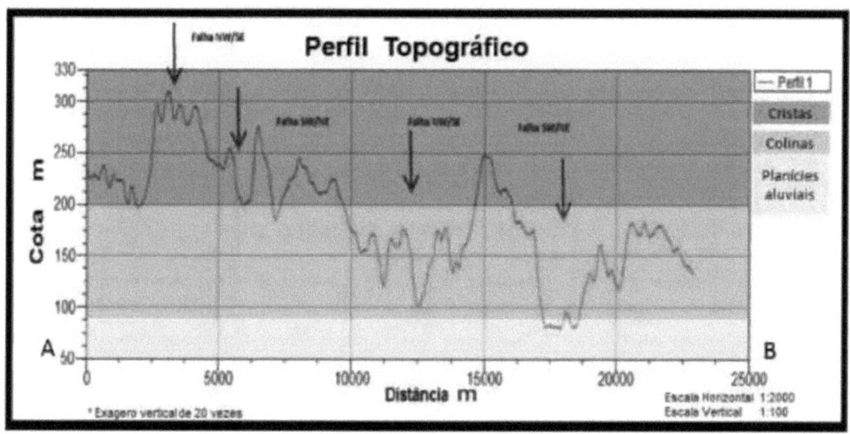

Figure 7: Topographic profile of *transect* A-B (map 4) in the northeast/southwest direction. Source: adapted from *Spring®*.

The crest compartment is predominantly in the northwestern region of the municipality, in the localities of Cerro dos Coqueiros, Sao Bràs Alto and Picada Grande, with elevations above 200 meters. The lowest elevations, between 40 and 80 meters, comprise the alluvial plains of Arroio Pinheiro and Sutil, to the west, and the alluvial plains of Arroio Duro, to the east of the municipality (map 4). The predominant slope class is from 8 to 13%, moderately undulating, with a greater occurrence of steep terrain to the northeast, exceeding 20% slope associated with crests.

The analysis of the control exerted by geomorphology on the development of erosion processes in the Municipality of Chuvisca was carried out taking into account the maps of relief units (map 4), the position in which these erosion features are located, the shapes of the terrain (vertical curvature and horizontal curvature), the slope of the terrain, ramp length and topographic factor (LS).

It was therefore found that the erosion processes mapped are located in the hill compartment or at the break in the relief, between crests and hills, therefore in hilly relief with slopes varying from moderately undulating (8%-13%) to undulating (13%-20%), where the process of relief dissection takes place.

When analyzing the representation of vertical and horizontal curvature results obtained from the SRTM (*Shuttle Radar Topographic Mission*) MNT data, it was found that 75% of the 32 registered gullies have a slope with vertical curvature in the profile in the classes: very concave to concave; and 59% have a slope with horizontal curvature in the plane in the classes: very convergent to convergent (tab. 1). These results coincide with those obtained by Viero (2004), in which 45% of the ravines and gullies studied in the Taboâo Basin, RS, were associated with slopes with vertical curvature in the profile as being concave and 58.3% with slopes with horizontal curvature in the plane as being convergent.

Table 1 - Percentage occurrence of gullies and gullies in each vertical and horizontal slope curvature class.

Curvature **Vertical in profile**	**%**	**Curvature** **Horizontal in the plane**	**%**
Very Concave	53,13	Very convergent	31,25
Concave	21,88	Convergent	28,13
Retilinea	6,24	Flat	18,75
Very convex	15,63	Very Divergent	9,37
Convex	3,12	Divergent	12,5

Source: adapted from MNT - SRTM (*Shuttle Radar Topographic Mission*).

According to Summerfield (1991), the shape of the slope has a significant effect on erosion, as it determines the path taken by the water, i.e. the distribution of the torrent over the terrain. According to the same author, laminar and subsurface flows are concentrated on converging contours, providing greater potential for furrow erosion. In contrast, laminar flow on slopes with divergent contours is dispersed downstream and erosion is minimized.

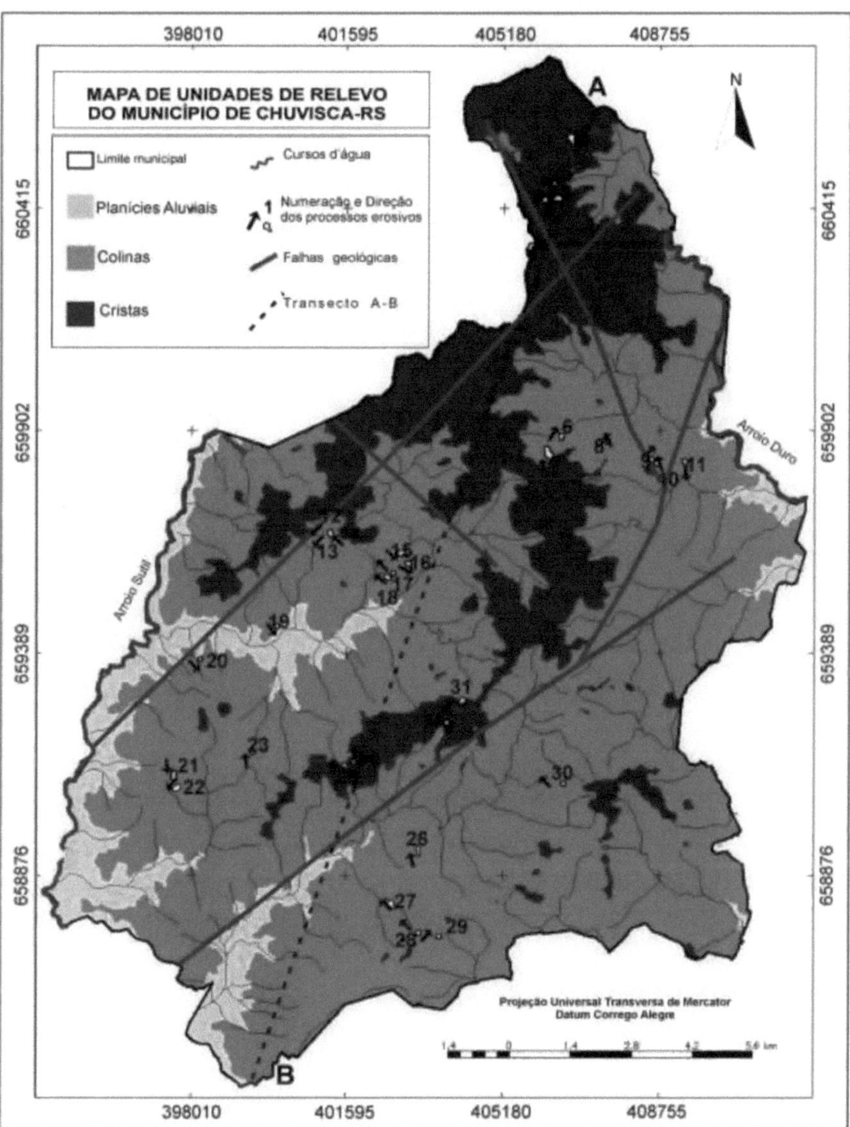

Map 4: Map of the Relief Units, showing the predominance of erosive processes in the hill and ridge compartments respectively, with an indication of the A-B *transect* (dotted line) used to represent the topographic profile. Source: prepared by the author, 2014.

The vertical and horizontal curvature classes, when combined and cross-referenced with the erosion points mapped, enabled 16 slope shapes to be indicated, as well as the observation of some of the characteristics of this variable. The largest number of gullies and gullies (18%) occurred on very concave and very convergent slopes and 15.63% on very concave and convergent slopes in Chuvisca (tab. 2). For Valeriano (2008), this would be one of the extreme cases of combining the curvature of the terrain with maximum concentration and accumulation of runoff. Another would be the convex-divergent shape, which provides maximum dispersion of runoff.

Table 2- Percentage of erosion and the shape of the slopes in the municipality of Chuvisca.

Vertical Curvature in Profile -	Horizontal Curvature in the Plane	Occurrences (%)
Very Còncava	Very convergent	18,75
Very Concave	Convergent	15,63
Very Concave	Gliding	12,50
Very convex	Very Divergent	6,25
Concave	Very convergent	6,25
Very Concave	Divergent	6,25
Concave	Convergent	6,25
Concave	Gliding	3,13
Very convex	Gliding	3,13
Concave	Very Divergent	3,13
Convex	Very convergent	3,13
Retilinea	Very convergent	3,13
Concave	Divergent	3,13
Very convex	Divergent	3,13
Very convex	Convergent	3,13
Retilinea	Convergent	3,13

Source: adapted from MNT - SRTM (*Shuttle Radar Topographic Mission*).

Therefore, we tried to identify in the field and with the help of GoogleEarth® images the existence of situations where runoff was concentrated. It was found that 69 % of the gullies and gullies mapped are located in a slope[1] with the presence of a convex-divergent slope upstream. The presence of convex-divergent slopes upstream provides a high dispersion of runoff into the talvegues, which are generally fragile points in the relief, due to the removal of

1 The line formed by the intersection of the two surfaces forming the slopes of a valley. It is the deepest part of the valley, where rainwater, rivers and streams flow. Source: Dicionario Livre de Geociências.

vegetation, excessive soil disturbance, poorly planned roads, the presence of an exposed saprolitic profile, among others.

The concentration of runoff in talvegues is natural and occurs due to the convergent shape of the slope. What's worse, of the 32 cases mapped, 22% occur in slopes with very close roads, providing an even more concentrated flow situation. In 50% of the ravines and gullies, there is a very close proximity (less than 1 meter) to roads, which are almost always located on the watershed. In other words, at the headwaters of erosion processes, artificially concentrating the surface flow, since in the cases recorded in the field, the rainwater drainage network of the roads is connected to the erosion points, strongly contributing to the instabilization of slopes and thus to the advance of erosion processes.

Another piece of data analyzed is the slope values of the terrain. At the points where the registered gullies and gullies occur, it was found that they develop over a wide range of slope values, varying from 3.2 to 30.4%, and that the largest number of these are between the classes: moderately undulating and undulating, which are the most common classes in the municipality (graph 1).

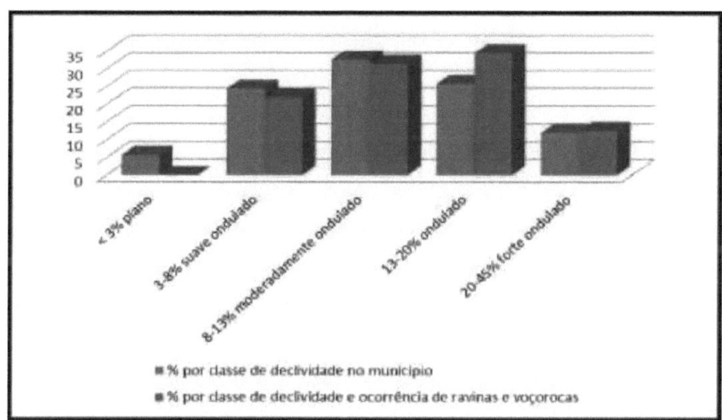

Graph 1: Graph showing the percentage of each slope class in the municipality and the occurrence of gullies and gullies. Source: adapted from MNT - SRTM (*Shuttle Radar Topographic Mission*).

On average, erosion occurs on land with a slope of 12.6% (moderately undulating). Due to the wide range of values, it is believed that the erosion features recorded occur mainly due to their association with other variables analyzed so far and not only due to the slope of the relief.

When analyzing the ramp length data from the points where gullies and gullies occur, it was found that 38% occur in terrain with a high ramp length (120 to 180 m) and that 34% occur associated with a ramp length of 120 to 180 m and a slope of 8 to 13%.

The topographic factor (LS), a relationship that includes the effects of the length of the slope (L) and the degree of slope (S), was obtained using the equation proposed by Wischmeier and Smith (1978) and divided into four classes: 0-1; 1-3; 35 and >5. There was a range of values from 0.2 to 21, with an average of 4.6 (class 3-5) (graph 2). As can be seen in graph 2, the highest number of erosion processes (40.6%) occurs in areas with an LS factor >5, despite the fact that class 1-3 predominates in the municipality.

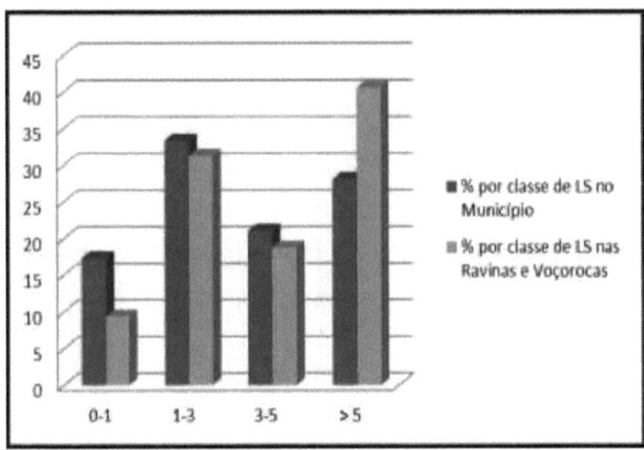

Graph 2: Frequency distribution of LS factor classes in the ravines and gullies identified in the municipality of Chuvisca, RS. Source: adapted from MNT - SRTM (*Shuttle Radar Topographic Mission*).

Although the topographical factor is important in defining the volume of runoff due to the increase in the catchment area (given by the length of the slope) and the speed of runoff (affected by the slope), the results indicate that this variable alone is not relevant in determining the occurrence of gullies and gullies in the municipality, since there was a very wide variation in values.

3.5. The pedological context as an indicator of erosion processes in the municipality of Chuvisca

As far as the soil is concerned, in the municipality of Chuvisca, according to the Soil Survey and Reconnaissance of the State of Rio Grande do Sul (IBGE, 1986), five mapping units are identified: Ca1 (Camaqua undulating relief), Ca2 (Camaqua strongly undulating relief), Cm (Cambai), PM1 (Pinheiro Machado undulating relief) and PM2-AR (Pinheiro Machado strongly undulating relief + Rock Outcrops). These units have been taxonomically updated by Streck *et al.* (2008) into umbric Argissolo Vermelho-Amarelo Distròfico (Camaqua), saprolitic Luvissolo Cròmico Pâlico (Cambai) and Neossolo Regolitico Distro-ùmbrico tipico ou léptico (Pinheiro Machado).

The soil map, based on IBGE (1986), shows three mapping units, unit PVd2, unit PVd13 and unit PBe1 (map 5). These units, updated according to EMBRAPA (2006), are represented by PVA1 - Argissolo Vermelho-Amarelo Distròfico abrûptico ou ùmbrico, PVA2 - Association Argissolo Vermelho-Amarelo Distròfico tipico ou ùmbrico ou Eutròfico tipico + Argissolo Bruno-Acinentado Alitico tipico + Neossolo Regolitico Distro- ùmbrico ou Distròfico léptico ou tipico ou Eutro-ùmbrico ou Eutròfico léptico ou tipico, and TXo - Luvissolo Hàplico Órtico tipico, equivalent to units PVd2, PVd13 and PBe1 from IBGE (1986), respectively. Of these, unit PVA2 predominates in the municipal area (90%), with unit TXo following (9%) and unit PVA1 a small part (1%). The PVA1 and TXo units form simple mapping units, with more than 70% of their area made up of just one type of soil, Argissolos and Luvissolos, respectively. In unit PVA2, there is no predominance of one type of soil over another, with Argissolos associated with Neossolos.

Of the 32 ravines and gullies registered in the municipality, 31 occur in the PVA2 unit made up of Argissolos associated with Neossolos and only one occurs in the TXo unit which corresponds to typical Luvissolos Hàplicos Órticos (map 5).

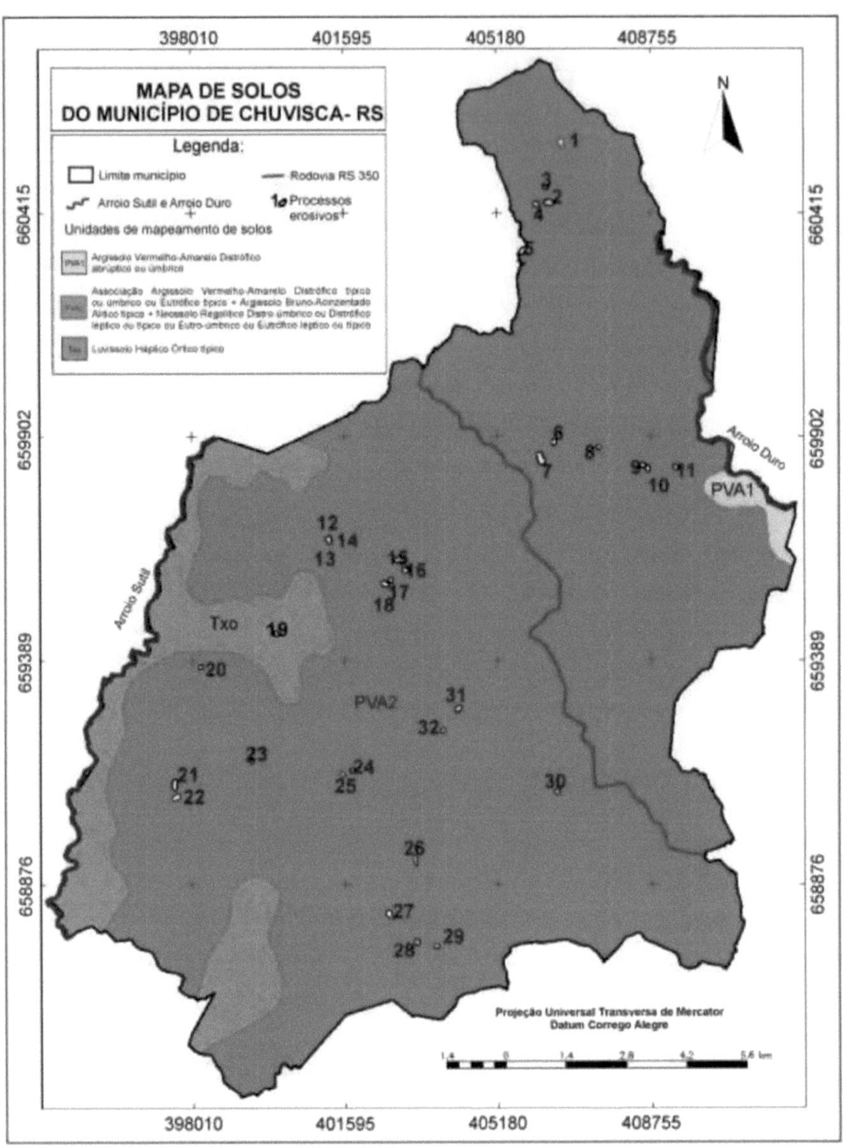

Map 5: Soil map of the municipality of Chuvisca, RS. Source: Adapted from IBGE (1986) and updated to the new EMBRAPA (2006) classification.

The term Argissolo derives from the presence of a more clayey subsurface horizon in the profile. According to Streck *et al. (2008),* Argisols are generally deep to very deep soils ranging from well-drained to imperfectly drained and usually have a textural Bt horizon. These soils can originate either from granites, as is the case in the municipality of Chuvisca, or from basalts, sandstones, claystones and siltstones. Textural horizon soils (Bt) are characterized by the presence of illuvial clay (migrated from the horizons above), which causes significant differences in the clay content between the A and B horizons (high textural ratio), moving from a sandier surface horizon to a clayier subsurface horizon. Although it may have relatively good aggregation in the surface horizons, this textural gradient makes it susceptible to erosion processes. Textural discontinuities are an obstacle to water infiltration along the profile, reducing its permeability and favoring surface runoff (STRECK *et al.*, 2008).

It should be noted that each type of soil has different physical attributes that impose varying erodibility on it. Among these, texture associated with particle size distribution is one of the most important factors. The authors Bertoni and Lombardi Neto (2005) and Guerra *et al.* (2005) agree with this statement and indicate soil structure, texture, permeability and density as important attributes for diagnosing susceptibility to erosion.

In addition to the abrupt textural change between the A and Bt horizons, Argissolos generally have the following aggravating factors in terms of their susceptibility to water erosion: (1) the high content of water-dispersed clay in the surface horizons, which indicates a greater amount of material available to be swept away in the downpour, and (2) the high content of easily erodible fine sand. An aggravating factor in Argissolos is the low percentage of clay in the other horizons, which, despite favoring water infiltration, does not contribute to good soil aggregation. The sand particles, in addition to providing a weak structure, are easily disaggregated and transported, making these soils more susceptible to water erosion.

Neosols, on the other hand, are young, undeveloped soils. Because they were formed recently from different types of rock, they usually have a sequence of two to three horizons. According to Streck *et al.* (2008), regolithic neosols are the most common type of neosol in Rio Grande do Sul and are characterized by their A horizon resting on completely altered rock.

3.6. Land use and property management practices as indicators of erosion processes in the municipality of Chuvisca

The first settlers in the south-central region, who had a tradition of farming, grew subsistence crops which, as they expanded, became a source of family income. When the municipality of Chuvisca was colonized, there was no monoculture; on the contrary, production was diversified and intense. According to the accounts of former residents, they lived off the consumption and sale of products such as corn, beans, wheat, linseed and sugar cane for the artisanal manufacture of cachaça. Firewood and meat were also sold in neighboring towns.

In the last three decades, however, the region has seen a rapid and vigorous occupation of its soil, with the arrival of tobacco companies financing tobacco production. Currently, the municipality's economy is based on planting and selling tobacco, with an annual production of 8,500 tons according to 2009 data from IBGE - Cidades@. Corn, cassava, beans and potatoes are also produced to a lesser extent (Table 3).

Table 3 - Data on the quantity produced and gross production value of the main temporary crops in the municipality of Chuvisca.

Cultivation	**Quantity produced (t)**	**Production value /thousand reais**
Smoke	8.805	37.651
Corn	10.000	3.399
Cassava	2.500	1.341
Beans	457	909
English potatoes	204	201

Source: IBGE - Temporary farming 2009

The land use classification generated for the municipality identified areas of native forest, afforestation, cultivation, exposed soil and fields (tab. 4 and map 6). There is a predominance of areas of exposed soil and forestation. The preponderance of exposed soil areas can be explained by the date of the image used for the classification - 3/10/2010 - corresponding to the time when soil preparation and tobacco planting activities are most intense.

Table 4: Calculation of areas by land use and cover class in the municipality of Chuvisca, RS.

Class	**Area (ha)**	**Area (%)**
Exposed Soil	7.954,02	36
Afforestation	4.918,95	22
Cultivation	4.007,34	18
Native Forest	3.788,46	17
Field	1.505,7	7
Total classified area	22.174,47	100%

Source: DUMMER, (2011).

The large number of forested areas is related to the expansion of eucalyptus (*Eucaliptus ssp.*) and black acacia (*Acacia mearnsi*) production in recent decades, to meet the need for firewood to dry the tobacco produced in the municipality. It should be remembered that it was difficult to classify these areas in the image, since the forestation appears in small plots that are mixed with areas of native forest. According to Collischonn (2009), because of the reforestation policy implemented by the agro-industry, characterized by individual negotiations with the integrated farmer, the plots of land used for forest cultivation are often half a hectare or even smaller.

For this reason, they are not distinguished by the shape or size of the plots in LANDSAT TM satellite images, since their size is usually between three and 10 *pixels* in the image (30mx30m or 20mx20m cells).

The cultivated areas also showed a significant increase, which is to be expected for a municipality that derives its income almost entirely from agriculture.

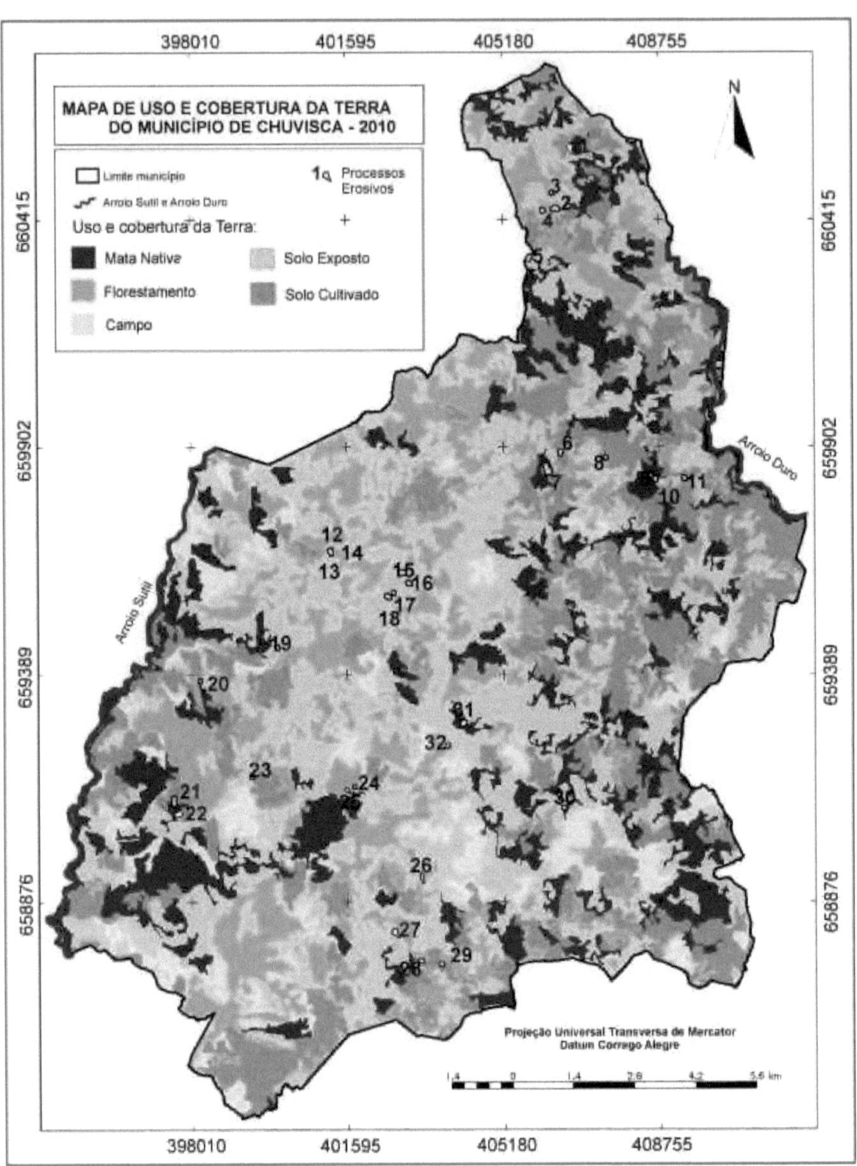

Map 6: Land use and cover map of the Municipality of Chuvisca, RS - 2010. Source: Adapted from DUMMER, J. (2011).

According to data on temporary crops in 2009 from IBGE - Cidades@, the municipality of Chuvisca has 4,200 hectares planted with tobacco, 4,000 hectares with maize and 520 hectares with beans.

However, there were few areas of fields and native vegetation. Due to inadequate soil management, since the arrival of the first settlers around 1900, cultivated areas have advanced under native vegetation, causing the rapid emergence of environmental problems such as soil degradation and erosion processes. The traditional opening up of new areas for cultivation ("roças"), since the arrival of the first farmers, has taken place through the cutting down of native species and the use of fires, a process which intensified after the 1980s, the period of greatest expansion of tobacco farming in the region. Without strict supervision, riparian zones were cleared of forest and crops were planted on the banks of streams and rivers, which had the direct consequence of silting up river channels, as was seen in the field (fig. 8). This exposure of the soil, especially in these areas of APP (Permanent Protection Areas), has paved the way for the development of erosion processes, such as those recorded in this study.

Figure 8: Watercourse with degraded riparian forest and agricultural plantations on the edge of the bed, Municipality of Chuvisca, RS. Photo: author's collection, 2013.

In turn, field studies have shown that 66% of the erosion processes recorded in the municipality of Chuvisca occur in areas where the soil has been conventionally prepared (plowing and harrowing) to grow tobacco, corn or artificial pasture. Field observations and interviews with 6 rural landowners, carried out between 2008 and 2014 at different times, show a strong influence of land use and cases of soil erosion, according to the registration forms (Annexes I, II, III, IV, V, VI).

There are situations in the municipality of Chuvisca where the geological formation, geomorphology and soil have the same characteristics, but the management differs, providing agricultural areas with gullies and gullies and agricultural areas without these erosive processes. In the town of Sao Bràs Alto, for example, two neighboring properties have the same characteristics in terms of geology, geomorphology and soil type, but what differs is their soil management. Both farms grow corn, pasture and tobacco. However, the first owner burns the corn straw and harrows the soil in order to plant tobacco without constructing contour lines and vegetated contour ridges. As an aggravating factor, the farmer makes intense use of heavy machinery, driving vertically down the slope, forming paths that over time become preferred points for surface runoff and the appearance of linear erosion (gullies and gullies) (figs. 9 and 10).

The farm has been used for more than 40 years for annual tobacco planting, using the conventional system. Conventional tobacco planting is carried out on a camalhao (murundus). According to Antoneli and Berdnaz (2010), the construction of this mound contributes to the formation of an ephemeral channel for rainwater runoff. The concentration of water between the rows (inter-soil) increases soil loss, mainly because the soil is constantly remobilized to eliminate weeds. Studies carried out by Antoneli and Berdnaz (2010), in Paranà, showed that soil losses in one harvest (September to February) amount to 27.5 t/ha, which is considered a high rate, as it refers only to the period when the soil was under tobacco cultivation.

Figura 9: Rural property in the town of Sâo Brâs Alto. The arrow in red indicates the ravine developed along the slope, and the arrows in yellow indicate the surface runoff from the lavoura directed towards the head and right side of the ravine. Photo: author's collection, 2013.

Figura 10: Linear erosion processes in a tobacco field under conventional planting - São Brâs Alto - Municipality of Chuvisca, RS. The arrows in red indicate the concentrated surface runoff forming erosion furrows along the field and the arrows in yellow indicate the flow towards the head of the ravine. Photo: author's collection, 2013.

In addition, the farmer does not properly organize surface runoff, directing rainwater to a weak point on the side of the field, where a ravine (6) has developed, according to map 2 and figure 11.

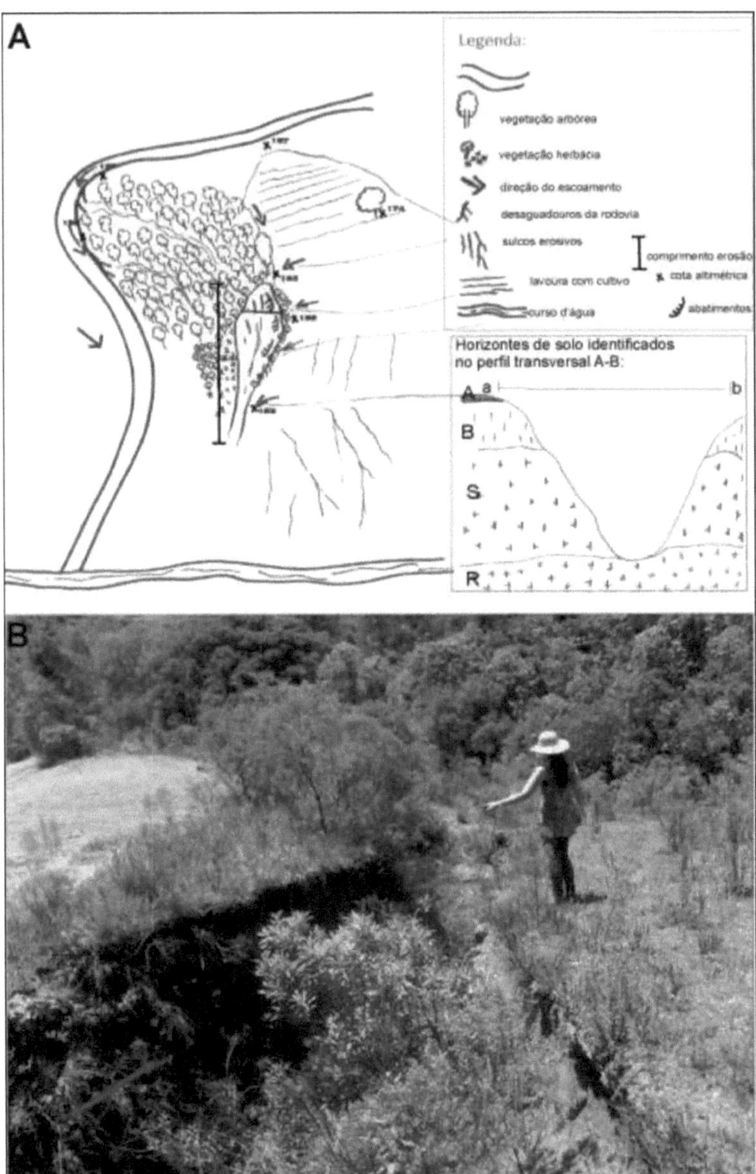

Figura 11: Ravine developed in a field on a rural property in the municipality of Chuvisca, RS. Sketch (A) showing the characteristics of the gully and its surroundings with a description of the lithological units identified in the cross-sectional profile A-B. Photograph (B) with arrow indicating the direction of erosion development. Org.: the author (2014).

According to Valentin *et al.* (2005), erosion processes in the form of gullies and gullies are most often triggered by inadequate cultivation practices and irrigation systems or overgrazing, road construction and urbanization. In this case, an interview with the owner revealed that the property has a history of degradation dating back some 80 years. Around 1940, the current owner's family started growing sugar cane in a rudimentary system, with intensive use of fire. Around 1980, the planting of tobacco, a new activity in the region, was interspersed with the sugar cane planting, until around the 1990s it became the only agricultural activity on the property, continuing to the present day.

The hypothesis that land use is one of the main factors conditioning erosion processes in the municipality is reaffirmed when analyzing the adjacent property, where there are no erosion processes. As well as keeping the slopes protected (fig. 12), the farmers practice no-till farming, with contour lines and vegetated contour lines, thus avoiding the concentration of runoff on a single point of the terrain and/or fragile areas.

Figure 12: Property with land parceling system (indicated by the yellow lines), contour lines (indicated by the black arrows) and vegetated slope (indicated by the red arrow). Photo: author's collection, 2014.

According to Filizola *et al.* (2011), the effectiveness of this practice is that it breaks down the energy of the runoff and deposits the sediment transported. Its great advantage is that it is easier to implement than terraces. As well as being an inexpensive practice, it helps to reduce erosion losses in both annual and perennial crops, since soil management is interspersed between one or two strips, and it is recommended to always start from the highest point of the slope, in a downstream direction (fig. 13).

Figure 13: Farming in parallel strips with different crops interspersed (delimitation indicated by the yellow lines). Photo: author's collection, 2014.

Regarding the occurrence of ravines and gullies in the municipality of Chuvisca and their relationship to land use, there was a significant occurrence (28%) of these processes in slopes on unpaved roads upstream. In many cases, in Permanent Preservation Areas (PPAs). It was found that the emergence of erosion processes, in many cases, in addition to being related to their position in the relief, also have roads at the headwaters of the main axes that have drains up to 30 meters long projected into their interior, as in one of the most prominent cases of linear erosion registered in the Municipality of Chuvisca (fig. 14).

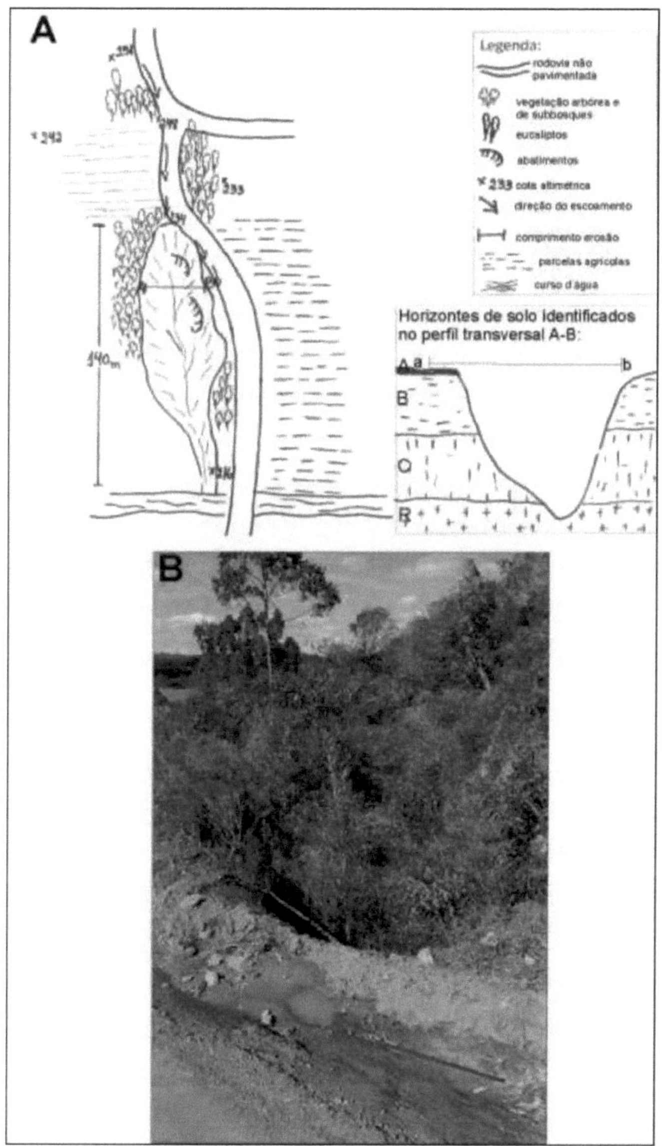

Figura 14: Gully in the town of São Brâs Alto, Chuvisca - RS. Sketch (A) showing the characteristics of the erosion and its surroundings and a description of the lithological units identified in the cross-sectional profile A-B. Photograph (B) showing the gully projected into the erosion (indicated by the arrow), the point of greatest instability. Org.: the author (2014).

The area occupied by erosion corresponds to part of a road and an old tobacco and corn farm, which has reached an intense stage of degradation and has been abandoned (2), according to map 2. The farm was abandoned about 20 years ago and today part of it and the road around it have been taken over by erosion, which has already reached the water table and is therefore a gully. Both the land to the south and on the north side (the edge of the road) belong to the same owners, who are losing land that can be used for planting, on the one hand due to the advance of erosion and on the other due to the advance of the road. As the gully continues to advance, mainly over the road, there is a need to correct its path, taking over productive areas of the rural property.

What can be observed is that in addition to the environmental factors, which are mainly related to the geological structure of the erosion and which follows the mapped lineaments, the inadequate use of the soil for agricultural production and the rainwater coming from the road could be the causes of the appearance of this gully.

The gully (1), according to map 2, located on the border between the municipality of Chuvisca and Dom Feliciano, has these characteristics (fig. 15). In an interview with the owner of the area, she said that the erosion currently at the edge of the road was located about 10m from it 40 years ago. The area had been cultivated with corn and had to be abandoned. Even after the introduction of Brachiaria-type vegetation and the installation of an electric fence to contain animal trampling on the right bank of the road, the gully is advancing towards the road and along secondary axes towards the crops (fig.15).

What can be seen is that, in addition to the characteristics of the environment showing indicators of fragility to erosion, the suppression of native forest, certainly carried out when the site was occupied, and the type of use and occupation of the land in agricultural practices, triggered the emergence of this gully. In addition, the reordering of the natural surface flow was modified when the side road was opened.

Similarly, erosion was mapped in the locality of Guaraxaim da Serra, no. 15 (according to map 2) and figure 16. This is an area with natural weaknesses related to the characteristics of the geological structure and geomorphology.

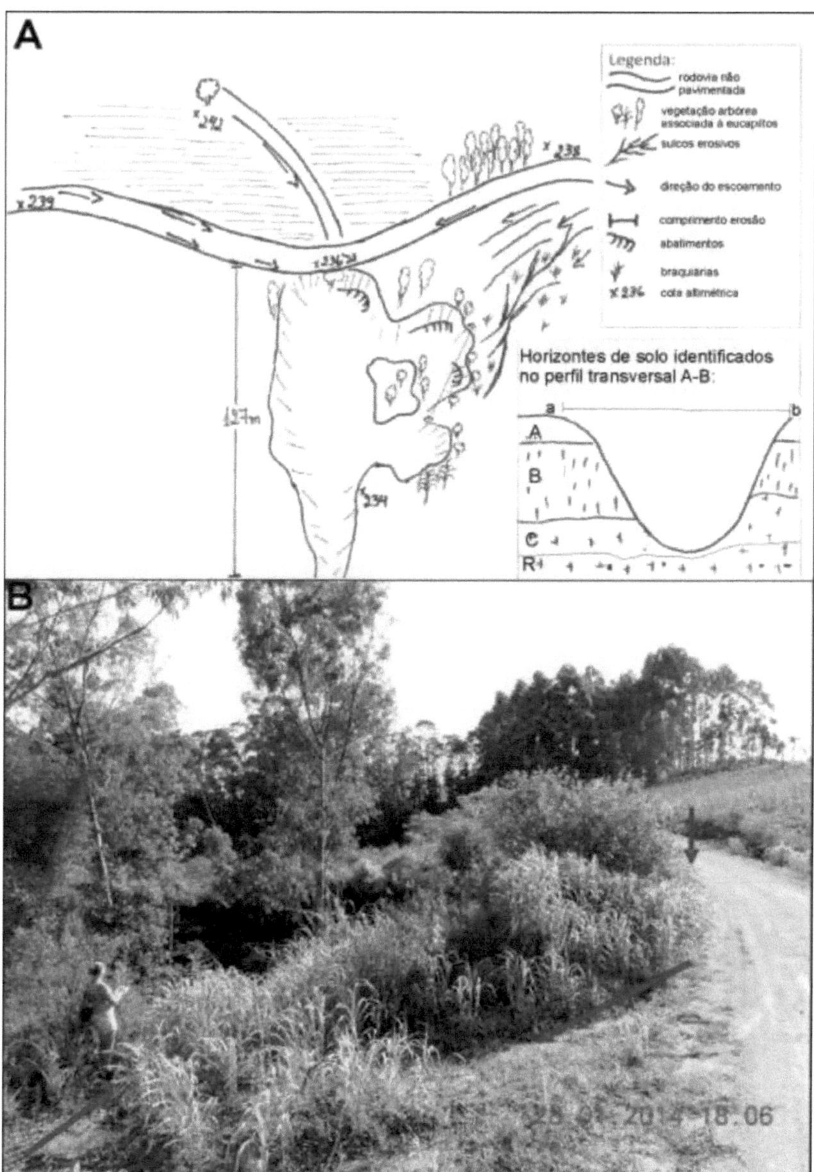

Figura 15: A gully developed in a cultivated area near an unpaved road in the town of Sâo Bràs Alto, Chuvisca-RS. Sketch (A), photograph (B) with arrows indicate the runoff in the surface drains (ditches) along the side road and which are concentrated at the head of the gully, in full stage of evolution. Org.: the author (2014).

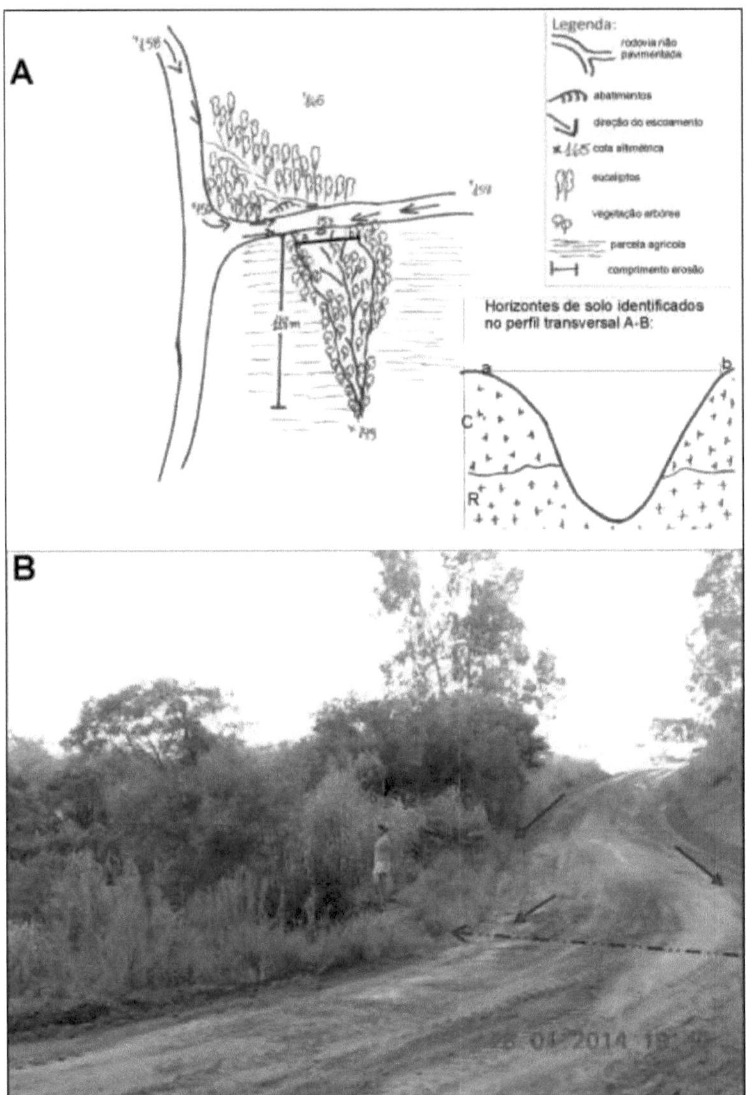

Figura 16: A gully developed in a cultivated area near an unpaved road, in the town of Guaraxaim da Serra, municipality of Chuvisca-RS. Sketch (A) shows the surface runoff directed towards erosion at various points on the slope, towards the head of the gully at a stage of widening and deepening of the main channel at base level, and exposure of the saprolitic horizon on the surface. The arrows indicate the runoff in the surface drains (ditches) and channeled subsurface drains (indicated by the dotted line) which are concentrated at the head of the gully, in the middle of its evolution. Org.: the author (2014).

The location and direction of the erosion coincide with the mapped lineaments and the divergent convex shape of the slopes. These characteristics provide a concentrated volume of rainfall runoff on the converging concave slope, where the erosion has developed. These natural weaknesses, which could be mitigated with proper planning and maintenance of the roads located upstream of the erosion, are accentuated by the presence of runoff channels directed at the erosion incision.

This situation is in line with other studies, such as the one carried out by Viero (2004), which have already shown that roads are responsible for the accumulation of surface flow due to the extensive channeling during downpours. Among the factors causing the higher volume of flooding due to roads is the compaction promoted in the terrain of the unpaved road, which reduces the infiltration rates of rainwater and increases the volume of surface runoff flowing into the drains directed to the side ditches of the side roads.

Similarly, roads that intersect streams receive and redirect the natural flow of runoff. In addition, the existence of a few drainage points along the route of these roads prevents the volume of runoff from being fractionated and makes the water catchment area at these points larger. The concentration of a volume of water in excess of the soil's infiltration capacity at the drainage points can lead to the emergence of gullies and gullies on the sides of side roads.

The remaining 6% of recorded erosion processes are in areas of natural grassland, which is used for livestock farming. In some way, they prove human action, both in causing and enhancing erosion. Two cases are worth highlighting. The first concerns a linear erosion located in a field area (21), according to map 2 and figure 17. In this erosion, it was possible to see that the water table had already been reached and that it was therefore a gully, which develops in Argissolos with a textural B horizon, which in turn is an obstacle to surface infiltration. According to the owner, the erosion has existed there for at least 28 years, when he bought the area. According to the owner, the site used to be a field and had a small ditch approximately 1m wide and 1m deep.

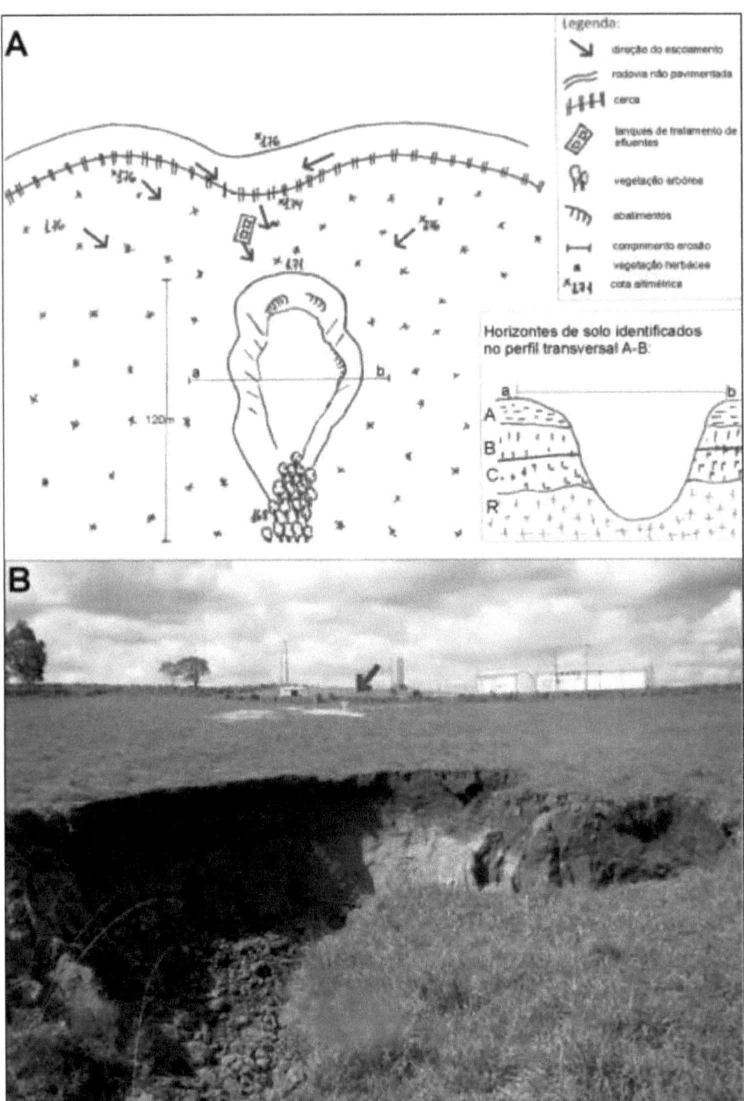

Figure 17: Gully in the town of Costa do Sutil, municipality of Chuvisca, RS. Sketch (top) showing the extent of the contributing area and the direction of the surface flow to the headwaters of the gully. Photograph (bottom) with arrow indicating the location of the mineral water industry's effluent treatment plant. Inside the gully we can see the A horizon in the process of being saturated with water by surface runoff. Org.: the author (2014).

There was a period when the field was used for planting corn, without the use of contour lines, which may have contributed to increasing the erosion process. Currently, the area is covered in herbaceous vegetation, and the site is used for livestock farming, where the contribution of animal trampling to the collapse of unstable slopes is visible (fig. 18).

Figure 18: Gully in the locality of Costa do Sutil, Municipality of Chuvisca. In detail, the trails made by trampling animals, which contribute to concentrated runoff into the gully. Photo: author's collection, 2014.

The instability of the gully slopes, especially upstream of the erosion incision, may be related to the presence of an effluent treatment plant from a mineral water industry (owned by the same owner), located upstream of the erosion (fig. 17). Indications of shrub and tree vegetation inside the gully and the presence of water saturation on its walls indicate that industrial effluent is gradually being discharged upstream. This would partly explain the progress of the erosion process in the upper headwaters, with the slope becoming undermined.

It can be seen that the erosion incision is disconnected from the drainage network, however, the road upstream and the concave convergent shape of the relief provide a concentration of rainfall runoff in rainy events. This indicates that the lack of organization of natural surface flow during rainy events, artificial flow due to effluent discharge and animal trampling on the edge of the gully are currently the agents accelerating the erosion process.

Another erosion process to be highlighted is in an area that is currently used as a field for animal husbandry, but which is nevertheless an area of suppressed native forest. The current resident (who has owned the property for more than 10 years) said that when he bought the property, the erosion process was already in place (32), map 2 (fig 19). According to him, erosion has been increasing for at least four years, mainly on a secondary lateral axis that runs towards the property's buildings. The same area visited four years ago was more stable than it is today, indicating that throughout history this land use could have been altering the water flow and accelerating the flow towards the main channel, causing it to become vertical.

In 2010, upstream of the area, it was not fully cultivated, but is now surrounded by cornfields, and the areas that used to be fields are now exposed soil for planting. Another aggravating factor is the abandoned road for vehicular traffic, with a large number of erosion furrows formed by concentrated runoff from the main road across it. The lack of drainage channels on the upstream roads, on a slope with an elevation of approximately 220 meters, generates a large volume of water runoff. This situation was confirmed by the heavy waterlogging of the soil within a 10-meter radius of the erosion, even four days after the end of the rain.

In order to summarize the main relationships identified between the 32 mapped features, with regard to environmental parameters such as lithology, lineaments, pedology (soil type) and geomorphology (slope shape), Table 5 was drawn up. An analysis of table 5 shows the strong relationship between erosion processes and the existence of regional lineaments in a northwest/southeast and northeast/southwest direction and the presence of Argissolos, since these two variables are the only ones present in the 4 relationships found.

Table 6 summarizes the main relationships identified between the mapped features in terms of land use (social) parameters, i.e. past vegetation cover and current use. This table shows the strong relationship between erosion processes and areas where natural forest has been suppressed and which are currently being misused, whether through agricultural and livestock activities or the lack of proper road maintenance.

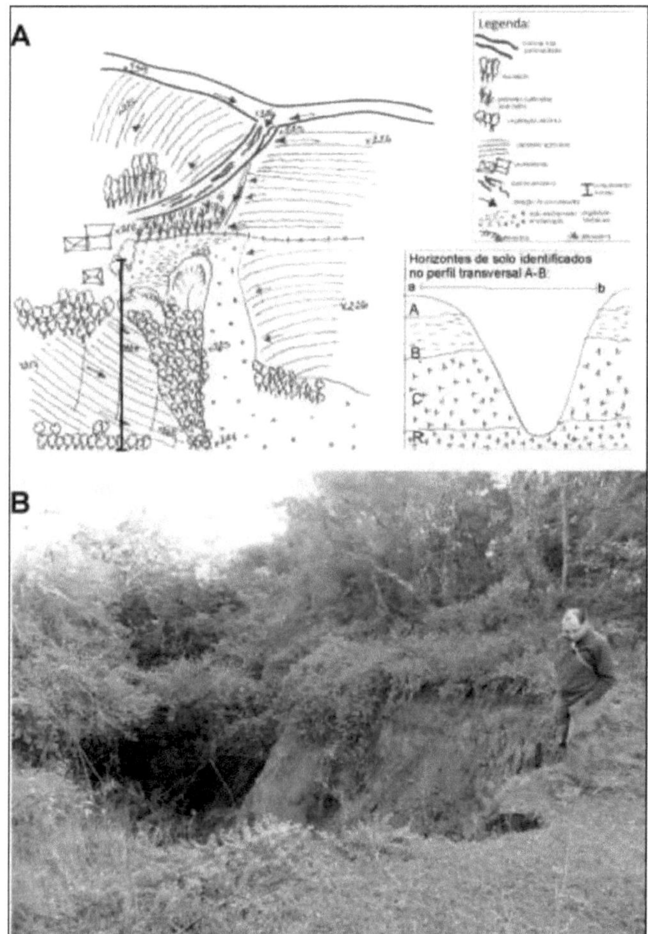

Figure 19: Ravine in a field area (APP), in the locality of Costa do Pinheiro, Municipality of Chuvisca-RS. Sketch (A) showing the characteristics of the ravine and its surroundings. The ravine has an extensive catchment area with surface flow (indicated by the arrows) directed towards it. In the photograph (B), the secondary axis is advancing over the property's improvements in the form of remontant erosion. Photo: author's collection, 2014.

Table 5: Summary of the relationships identified with regard to mapped erosion and environmental parameters.

Lithology	**Lineaments**	**Type of soil**	**Geomorphology (slope shape)**
Suite D Feliciano Cerro Grande	Northwest/Southeast Northeast/Southwest	Argisols associated with Neosols	M. Convex and M. Divergent.
Suite D Feliciano Cerro	Northwest/Southeast	Argisols associated with	M. Còncava e Còncava e M.

Grande	Northeast/Southwest	Neosols	Convergente a Convergente
Pinheiro Machado	Northwest/Southeast Northeast/Southwest	Argisols associated with Neosols	M. Concave to concave and M. Convergent to convergent
Pinheiro Machado	Northwest/Southeast Northeast/Southwest	Argisols associated with Neosols	M. Convex to convex and M. Convergent to convergent

Source: The author (2014).

Table 6: Summary of the relationships identified between mapped erosion and land use parameters (social).

Vegetable topping (pretèrita)*	**Current use**
Semideciduous Seasonal Forest	Degraded vegetation/shrub cover
Semideciduous Seasonal Forest	Field used for livestock
Semideciduous Seasonal Forest	Conventional cultivation
Semideciduous Seasonal Forest	Unpaved roads

*Source: Levantamento de Recursos Naturais V. 33, IBGE. Org.: the author (2014).

This confirms that all the erosion processes coincide with the lineaments and the presence of Argissolos (the class with the highest occurrence in the municipality) on fragile slopes due to the suppression of natural vegetation and exposure resulting from different agricultural uses and practices. Thus, at least three types of erosion processes were identified in the municipality: (i) gullies and gullies developed in areas of fields used for livestock farming; (ii) gullies and gullies developed in crop areas; and (iii) gullies and gullies developed on the sides of roads. Some examples of the types of ravines and gullies mapped can be seen in figure 20.

Figure 20: Some of the ravines and gullies recorded in different localities in the Municipality of Chuvisca by type of erosion process. I) erosion in a field area - Costa do Sutil locality (21); II) erosion in a nearby gully area - Costa de Sâo Bràs Alto locality (31); III) erosion on a road - Sâo Bràs locality (24). Org.: the author (2014).

6. FINAL CONSIDERATIONS

Initially, through the existing register of 32 gullies/voçorocas in the municipality of Chuvisca, this research enabled an analysis of the environmental conditions in relation to the emergence of these erosive processes. With regard to the geological-pedological constraints, it was found that most of the gullies and gullies develop on a highly weathered rocky substrate, which indicates a natural fragility, especially when the soil is subjected to the concentration of surface runoff and the saprolite does not have sufficient structural stability to contain the erosive processes. It was found that all the ravines and gullies mapped in the municipality follow the direction of the set of tectonic faults developed in the main direction NE/SO and the secondary direction NO/SE. This proves that, in many cases, erosion processes are influenced by local/regional tectonics, both in their location on the ground and in their evolution, since it was proven that the development of the mapped ravines and gullies follows the preferential direction of the lineaments and faults.

The analysis of the municipality's soil types and the emergence of the mapped gullies and gullies showed, based on the soil classification available for the region, that 99% of the erosion processes develop in Abrupt Red-Yellow Dystrophic Argisols, characterized by abrupt textural change, a class that predominates in the region, occupying 90% of the municipality's territory. The presence of a subsurface horizon with a concentration of clay makes this soil more susceptible to erosion, as it hinders the infiltration of water into the subsurface, thus facilitating surface runoff.

With regard to the geomorphological conditions and the development of gullies and gullies in the municipality of Chuvisca, it was found that 34% of the recorded erosion occurred on undulating terrain with slopes between 13 and 20%. The highest proportion of gullies (38 %) occurred on land with slope lengths of between 120 and 180 meters, and 31 % in areas with a topographic factor greater than 5%. However, due to the wide range of values, these variables, if analyzed in isolation, were not relevant enough for the emergence of the erosion processes mapped in the municipality.

It was also observed that 69 % of the gullies and gullies were in streams with the presence of diverging convex slopes upstream, leading us to believe that this condition is important in increasing the chances of forming such erosive features. In this situation, especially if associated with high slopes and ramp lengths, the potential kinetic energy of rain is high and the risks of erosion become even greater.

The geomorphology of the terrain proved to be a relevant variable in the probability of linear erosion occurring, confirming that most of the erosion processes recorded are located in the crest or hill compartments, in a natural process of relief dissection in relation to the regional base level. Allied to this condition, the fact that most of the ravines and gullies develop on slopes with the association of divergent convex terrain upstream, contributes to increasing the potential risk of ravines and gullies forming on concave and convergent points of the terrain. Of the 32 recorded occurrences, 75% are located on terrain with surfaces with a very concave to concave profile; 59% occur on surfaces with a very convergent to convergent plane, these surfaces having the highest concentration of surface flow.

The analysis of the control exerted by land use and agricultural practices on the emergence of erosion processes in the municipality showed this to be the most important variable, since situations were found in which the conditioning factors of the environment presented the same characteristics in erosive morphologies, however, there are properties that present several erosive processes and properties that do not, showing that different land uses are one of the factors that trigger or mitigate erosion. Of the ravines and gullies registered and analyzed, 66% occur in areas of corn and tobacco crops, under conventional planting; 28% are located in Permanent Preservation Areas (PPAs), which have been degraded over time by the suppression of native forest, mainly due to the advance of roads and crops, with 6% in areas of fields used for livestock. Of the cases registered, 50% are destabilizing roads and constituting risk areas for the population, as well as contributing to higher maintenance costs for unpaved roads every rainy season.

Finally, the study is expected to continue and deepen, considering that the linear erosion processes recorded would have their genesis related to two more hypotheses, since in some cases a lowering of the base level was found. The first hypothesis would be related to the greater amount of water entering the system, which in turn would be responsible for the acceleration of concentrated runoff processes, and thus the verticalization of the drainage network. The second would be related to the influence of neotectonics which could be slowly uplifting the Shield, activating lineaments and faults and accelerating the notching of the channels due to the increase in potential energy and incision by the water.

REFERENCES

ANTONELI, V. and BEDNARZ J. A. **Soil erosion under the cultivation of tobacco (*nicotina tabacun*) on a small rural property in the municipality of Irati** - Paranà. CAMINHOS DE GEOGRAFIA - online journal http://www.ig.ufu.br/revista/caminhos.html ISSN 1678-6343. 18p.

BAHIA, V. G.; CURI, N.; CARMO, D. N. Fundamentals of soil erosion. **Informe Agropecuàrio**, Belo Horizonte, v. 16, n. 176, p. 25-31, 1992.

BEAVIS, S.G. **Structural controls on the orientation of erosion gullies in midwestern New South Wales, Australia**. Geomorphology, v. 33, p. 59-72. 2000

BERTONI, José; LOMBARDI, Francisco Neto. 5.ed. **Soil conservation**. Sao Paulo: icone, 2005. 355 p.

BOIFFIN, J et BRESSON, L.M. Dynamique Deformation des Croûtes superficielles: apport de l'analyse microscopique. Paris AISS/AFES 1987.

CAMARA, G.; SOUZA, R.C.M.; FREITAS U.M, GARRIDO J. "SPRING®: Integrating remote sensingand GIS by object-oriented data modelling" Computers & Graphics, v.20:, n.3, p.395-403, may-jun 1996.

CAREY, B. **Erosion Gully**. 2006. Available at: < http://www.derm.qld.gov.au/factsheets/pdf/land/l81.pdf) >. Accessed on: November 19, 2013.

CASSETI, Valter. **Geomorphology**. [S.l.]: [2005]. Available

at: <http://www.funape.org.br/geomorfologia/>. Accessed on: June 31, 2014.

CASSOL, Elemar. Antônio; LIMA, V. S. **Erosion between furrows under different types of soil preparation and management. Pesquisa Agropecuària**. Brasileira, Brasilia, v. 38, n. 1, 2003, p. 117-124.

CASSOL, Elemar. Antônio. **Soil erosion**. Porto Alegre: UFRGS, Faculty of Agronomy, Department of Soils, 2007. 20 p. Class notes for AGR 03006 - Soil Erosion and Conservation.

CHEMALE Jr. F. **Evoluçao Geológica do Escudo Sul-rio-grandense**. In Hoz, M and L. F. de Ros (Eds.) Geology of Rio Grande do Sul. Porto Alegre. Centro de Investigaçao de gonwana/Universidade Federal do Rio Grande do Sul, p. 13-52.

COLLISCHONN, Erika. 2009. **Flooding in Venâncio Aires/RS: Interactions between natural and social dynamics in the formation of urban socio-environmental risks**.

2009 (PhD Thesis in Geography). Federal University of Santa Catarina.

CPRM. **Geological Map of Southern Coverage, scale 1:250.000**. Brasilia: Brazilian Geological Service, 2003. CD-Rom

CPRM. **Geological Map of Rio Grande do Sul, scale 1:750.000**. Brasilia: Brazilian Geological Survey, 2007. CD-Rom.

DAEE - Department of Water and Electricity. **Erosion control:** conceptual and technical bases; guidelines for urban and regional planning; guidelines for the control of urban gullies. Sao Paulo: DAEE/IPT, 1989. 92p.

DENARDIN, José Eloir *et al*. **Rainfall management in no-till systems**. Porto Alegre: State Forum on Soil and Water, 2005. 88p

DUMMER, Juliana. Voçorocas no Meio Rural: **Um Diagnóstico de Processos Erosivos no Municipio de Chuvisca, RS**. 112 f. 2011. Monograph (Bachelor's Degree in Geography). Institute of Human Sciences - Federal University of Pelotas.

DUMMER, Juliana. ; KOESTER, Edinei; BRUCH, Alexandre. F. **Geological Survey to Study Soil Erosion in the Municipality of Chuvisca, RS**. Proceedings of the XVI National Meeting of Geography .n°1296. Porto Alegre, 2010.

ELLISON, W. D. **Soil erosion studies**. Agric. Eng., St. Joseph, v. 28, 1947.

EMBRAPA - BRAZILIAN AGRICULTURAL RESEARCH COMPANY. National Soil Research Center (Rio de Janeiro). **Brazilian Soil Classification System**. 2. ed. Rio de Janeiro: EMBRAPA Solos, 2006. 306p.

FILIZOLA et al. **Control of Linear Erosion Processes (gullies and gullies) in Sandy Soil Areas**. Technical Circular 22. Embrapa. Jariuna, Sao Paulo. 2011. Available at: http://www.cnpma.embrapa.br/download /circular_22.pdf. Accessed on: July 13, 2014.

GUERRA, Antônio, José. Teixeira; SILVA, Antônio. Soares. **Erosion and Soil Conservation:** applications, themes and concepts. RJ: Bertrand Brasil, 2005. 340p.

GUERRA, Antônio. José. Teixeira; ARAUJO, Gustavo. H. de S; ALMEIDA, J. R. **Gestâo Ambiental de Areas Degradadas**. 2. ed. Rio de Janeiro: Bertrand, 2007. 320p.

HASENACK, H.; WEBER, E. J. **Continuous vector mapping of Rio Grande do Sul on a scale of 1: 50,000**. Porto Alegre, RS: UFRGS IB Ecology Center, 2010. DVD.

BRAZILIAN INSTITUTE OF GEOGRAPHY AND STATISTICS. **Survey of Natural Resources**. Rio de Janeiro: IBGE, 1986. 796p, v.33.

BRAZILIAN INSTITUTE OF GEOGRAPHY AND STATISTICS - **Cities**. 2009 Agricultural Census. Available at < http://www.ibge.gov.br/cidadesat/topwind ow.htm?1>. Accessed on: January 10, 2014.

MENDONÇA

MALIK, N. Javed; MOHANTY C. **Active tectonic influence on the evolution of drainage and landscape: Geomorphic signatures from frontal and hinterland areas along the Northwestern Himalaya, India**. Journal of Asian Earth Sciences 29 (2007) 604-618. 28 March 2006. Available at: www.elsevier.com/locate/jaes. Accessed on August 10, 2014

MEYER, L. D.; FOSTER, G. R.; NIKOLOW, S. **Effect of flow rate and canopy on rill erosion**. Trans. Of the ASAE, St. Joseph, v. 18, n. 5, 1975.

PHILIPP, Rui Paulo; NARDI, Lauro Valentin Stoll; BITENCOURT, Maria de Fâtima. **The Pelotas batholith in Rio Grande do Sul**. In: HOLZ, Michael, ROS, Luiz Fernando De (EDS). Geology of Rio Grande do Sul. Porto Alegre: CIGO/UFRGS Porto Alegre, 2000, p. 133-160

POUSEN, Jean. **Gully Typoloy na gully control measures in the European loess belt**. Soil Erosion Agricultural Land. Leuven, 1993. P. 513-530.

PRESS, Frank; SIEVER, Raymond; GROTZINGER, John, JORDAN, Thomas H. **To Understand the Earth**. Porto Alegre, Editora Artmed, 2006. 656p.

OLIVEIRA, M.A.T. Erosion processes and preservation of areas at risk of gully erosion. In. GUERRA, A. J. T; SILVA, A. S; BOTELHO, R.G. M. (eds.). Soil erosion and conservation: concepts, themes and applications. Rio de Janeiro: Bertrand Brasil. 1999. p. 57-99.

RAMALHO, Maria Francisca de Jesus. **Geomorophology and Environmental Dynamics: Pitimbu river valley**. Natal, RN. Imagem Gràfica, 2003. 87p.

REZENDE, M.; CURI, N.: REZENDE, S. B.; CORRÊA, G.F. **Pedologia: base para distinçâo de ambientes**. Ed. Viçosa: NEPUT, 1997. 367p.

SANCHEZ Rodrigo Baracat *et al*. **Spatial variability of *soil* attributes and erosion factors in different pedoforms**. Revista Brasileira Scilelo Solo, Bragantia, Campinas, v.68, n.4, p.1095-1103, 2009. Available at: http://www.scielo.br/pdf/brag/v68n4/ v68 n4a30.pdf. Accessed on: July 28, 2014.

SANTOS, H. G. et al. **Sistema Brasileiro de Classificaçâo** de Solos. 2 ed. Rio de Janeiro: Embrapa Solos, 2006. 306p.

STRECK, Edemar Valdir et al. **Soils of Rio Grande do Sul**. 2. Ed. Porto Alegre:

EMATER/RS, 2008. 22p.

SÂO PAULO. Department of Energy and Sanitation. Department of Water and Electricity. **Erosion Control: conceptual and technical bases; guidelines for urban and regional planning; orientation for the control of urban gullies**. 2.ed. Sao Paulo, DAEE/IPT, 1990. 92p.

SILVA, Tiago Pinto da *et al.* **A Influência de Aspectos Geológicos na Erosao Linear - mèdio baixo vale do Ribeirao do Secretàrio, Paty do Alferes (RJ)**. Geosul, Florianópolis, v. 18, n. 36, p 131-150, jul./dez. 2003.

SUMMERFIELD, M. A. **Global Geomorphology: na introduction to the study of landforms**. New York: Longman Scientific & Technical, 1991. 537 p: ill.

TRICART, J. **Ecodynamics**. IBGE. Rio de Janeiro. Brazil, 1977.

VALERIANO, M. de M. **Topodata guia para utilizaçao** de **dados Geomorfológicos locais**. National Institute for Space Research (INPE) Sao José dos Campos 73p. 2008. Available at: <. http://www.dsr.inpe.br/topodata/ data/guia_utilizacao_topodata.pdf.>. accessed on June 5, 2011.

VALENTIN C. J.; POUSEN, Y. L. **Gully Erosion: Impacts, factors and control**. Elsevier B.V. 2005. 132-153. Available at: www.elsevier.com/locate/catena. Accessed on: September 20, 2014.

VIERO, Ana Claudia. **Analysis of geology, geomorphology and soils in the gully erosion process: Taboao Basin, RS**. 129f Dissertation in Water Resources and Environmental Sanitation. Hydraulic Research Institute, UFRGS, Porto Alegre. 2004.

WANG, G.; FANG, S.; SHINKAVERA, S.; GERTNER, G.; ANDERSON, A. **Spatial uncertainty in prediction of the topographical factor for the resided universal soil loss equation (RUSLE)**. Transactions of the ASAE, v.45, p.109-118, 2002.

WISCHMEIER, W.H.; SMITH, D.D. **Predicting rainfall erosion losses:** a guide to conservation planning**.** Washington: United States Department of Agriculture, 1978. p 58.

ANNEX

ANNEX I: Registration form for linear erosion number 1 (map 2).

EROSION REGISTRATION FORM		
I. Identification and location of erosion	State: *RS*	Municipality: *Chuvisca*
Description: *Border gully.*		**Location:** *Property of Mr. Erno Peter*
Access: *Serro dos Coqueiros road,*		
II. Regional data		
Watershed: *Arroio Duro Watershed*		**Geomorphology:** *Ridge Compartment*
Geology: *Dom Feliciano Suite, Cerro Grande facies*		**Pedology:** *(PVA2), composed of Argissolos associated with Neossolos.*
Original vegetation: *Forest typical of APP removed to plant fields and pastures*		
III. Dimensions of erosion		
Approximate length: *126m*		
IV. Characteristics of the contributing area		
Use and occupation of the contributing area: *Pasture cultivated with brachiaria decumbens (Brachiaria decumbens) on the left and acacia woodland on the right. APP (downstream of the slope), unpaved road and crops upstream.*		
Causes, conditioning factors and mitigating factors: *The landowner who inherited the land, Mrs. Zilma, the wife of Mr. Erno Peter, has lived there for 68 years, and said that she remembers the erosion from her childhood. According to her, 40 years ago the erosion was located about 10m from the road. Corn has always been grown there using contour lines. According to Ms. Zilma, her parents were already fighting to contain the erosion by placing logs inside the pit and planting elephant grass and bamboo inside. More recently (in the last 10 years) the site was abandoned for corn plantations, brachiaria was introduced as a ground cover and an electric fence was installed to prevent cattle from accessing the erosion since, according to the owners, animals are eventually let loose on the site to graze. According to the landowner, the erosion is caused by rainwater on both sides of the road. About two years ago, after a request to the town hall, an attempt was made to divert the road, which, without maintenance, has contributed to stabilizing the erosion process. In summary, the characteristics of the local physical environment indicate the fragility of the area in terms of the development of large-scale linear erosion, due above all to the fact that the slope is surrounded by divergent convex slopes that naturally direct runoff to the point where the erosion developed. Combined with the suppression of native forest, certainly carried out when the site was occupied, and the type of use and occupation of the land for agricultural practices, this erosion has advanced to the stage of a gully. According to the landowner, there is an outcrop of water downstream of the main axis. Still in terms of land use, there is a lack of natural surface flow management during rainy events.*		
V. Characteristics of linear erosion		
Erosion **dynamics (general characteristics):** *the erosion has a coalescing form, with secondary axes that are quite advanced to the right, with a wider headland than the downstream side (fig.). The central axis is the most stable and the secondary axes are quite unstable, showing successive slumps on fault planes (slickensides). The slopes are steep throughout the erosion.*		
Evolution forecasts: *There are many furrows upstream of the axes due to concentrated surface runoff in arable land with low vegetation cover and a clumped root system, which facilitates the formation of concentrated runoff channels towards the lateral axes of the gully, contributing to the occurrence of localized mass movements such as successive subsidence and/or slope collapses. Also upstream, the left side of the main headland receives channeled runoff from the road on the right and left. These mechanisms give instability to this erosion, which tends to evolve again and again, increasing in length and laterally, due to future junctions of its minor axes.*		
VI. Main impacts		
The outbreak of this erosion triggered the following impacts: *- Local soil loss with reduction of pasture area;* *- Loss of usable area upstream due to the relocation of the road.* *- Risk of accidents due to falling animals and vehicles.*		
VII. Preventive and containment measures implemented:		
The area was abandoned for cultivation and plants of the Bachiaria type were introduced upstream of the unstable shafts, insulated with electric wire and both measures are currently unmaintained.		
VIII Suggested preventive and containment measures:		
As a preventative measure for the erosion process, it is suggested that local land use be adapted to its carrying capacity, avoiding the triggering of environmental impacts of this nature, while the definitive containment alternatives aim to stabilize the erosion occurrence through vegetation practices (native and exotic plants) up to the implementation of more complex engineering and bioengineering works. *Among the many possible interventions to stabilize the erosion, the following are suggested: a) Complete isolation of the area within a radius of 10 meters or more from the edges of the erosion, to prevent animal traffic (cattle and horses) and encourage spontaneous repopulation around the erosion;* *b) Implement contour lines and native undergrowth typical of the region in the lateral contribution area to reduce the speed of surface runoff that reaches the most unstable lateral axes;* *c) upstream at the headwaters reorganize rainwater runoff with a greater number of properly planned and vegetated drains, adopting contour lines in crops located upstream beyond the road, avoiding the collapse of slopes and erosion.*		
IX. Identification of the form		
References: Mr. Wegner		
Team: *Juliana Dummer (master's student); Dr. Roberto Verdum*	Location: *30°40 31.66"S*	Date:

(advisor)	*51°58'28.42'"O51°58'28.42'"* Altitude: *236m*	*20/06/2014*

ANNEX II: Linear erosion record sheet number 2 (map 2).

EROSION REGISTRATION FORM	
I. Identification and location of erosion State: RS Municipality: Chuvisca	
Description: Tietz gully	**Location:** Prop. Mr. Nilton Tietz
Access: Serro dos Coqueiros road,	
II. Regional data	
Watershed: Arroio Duro Watershed	**Geomorphology: Ridge Compartment**
Geology: *Pinheiro Machado Complex composed of whitish gray granodiorites, porphyritic gray granodiorites and veins of pink granodiorites measuring 30 cm thick.*	**Pedology:** *(PVA2), composed of Argissolos associated with Neossolos.*
Original Vegelation:	
III. Dimensions of erosion	
Length: 140m	
IV. Characteristics of the contribution area	
Use and occupation of the area of contribution: unpaved road, tobacco plantation abandoned 20 years ago, currently cattle-ranching.	
Causes, conditioning factors and mitigating factors: *A former resident of the area said that this erosion has existed for at least 30 years. According to the resident, the area currently occupied by the erosion is part of a field that was abandoned about 20 years ago, which has reached an intense stage of degradation and has been abandoned, along with part of a road. As erosion continues, it is mainly on the road. There is a need to correct its route, taking over productive areas also owned by the same landowner. In the area above the capoeirão, secondary forest was identified with a greater amount of leaf litter. This area is indicative of the original forest that existed both in the abandoned farmland (the area under the capoeirão) and in the gully, which was possibly cleared when it was occupied for agricultural purposes. The genesis of the erosion is associated not only with the inappropriate use of the soil for agricultural production, but mainly with the rainwater coming from the road, which is concentrated at a single point upstream of the erosion incision. Both the land to the south and to the north (the edge of the road) belong to the same owners, who are losing land that can be used for planting, on the one hand due to the advance of erosion, and on the other due to the advance of the road. Erosion has already reached the water table, so this is a gully with more stable walls to the south, possibly due to the presence of vegetation in the area above, which belongs to part of the abandoned farmland with vegetation cover composed mainly of pioneer species.*	
V. Characteristics of linear erosion	
Erosion **dynamics (general characteristics): the** *gully develops along a single axis, with subsidence on the right-hand side (less stable) bordering the highway due to the concentration of ditch runoff. The slopes slope steeply along the entire stretch of erosion, with the slopes on the left showing greater stability due to the tree vegetation that has developed since the crop was abandoned around 20 years ago.*	
Erosion forecasts: *Due to the presence of rainwater run-off channels at the headwaters and on the right-hand side of the erosion, it is expected that the erosion will advance over the road laterally and in depth up to the base level of the watercourse downstream.*	
VI. Main impacts	
Its impact is related to the advance on the access road to the locality, as well as the loss of productive agricultural land, silting up of drains and safety risks for the population living in its surroundings, since there have already been cases of accidents involving vehicles and animals at the site.	
VII. Preventive and containment measures implemented:	
There are no preventative or containment measures in place.	
VIII Suggested preventive and containment measures:	
Isolate the area and remove the flow of rainwater from inside the gully, fractionating it by installing a greater number of properly planned drainage channels along the road; In dry periods, smooth out the slopes of the gullies with the help of machinery, minimizing the risk of collapse by verticalizing the walls; Install containment barriers inside the gullies, with material available on the rural property (bamboo poles or eucalyptus stakes) tied together with ramie twine or vines. Place drains of biodegradable material such as bidim or burlap in the containment barriers.	
IX. Identification of the form	
References: *Mr. Nilton Tietz*	
Team: *Juliana Dummer (master's student); Dr. Roberto Verdum (advisor)*	Location: *30°41· 21.59" 51°58'41.15"O* SDate: *18/11/2012* Altitude: *226m*

ANNEX III: Linear erosion record sheet number 6 (map 2).

<table>
<tr><td colspan="3">EROSION REGISTRATION FORM</td></tr>
<tr><td colspan="3">I. Identification and location of erosion State: RS Municipality: Chuvisca</td></tr>
<tr><td>Description: Mr. Josino's Farm Ravine</td><td colspan="2">Location: Sâo Bràs Mèdio, Prop. Of Mr. Josino</td></tr>
<tr><td colspan="3">Access: Vino Peter Road</td></tr>
<tr><td colspan="3">II. Regional data</td></tr>
<tr><td>Hydrographic Basin: Hydrographic Basin of the Arroio Duro</td><td colspan="2">Geomorphology: Hill Compartment</td></tr>
<tr><td>Geology: Pinheiro Machado Complex composed of whitish gray granodiorites, porphyritic gray granodiorites and veins of pink granodiorites measuring 30 cm thick.</td><td colspan="2">Pedology: (PVA2), composed of Argissolos associated with Neossolos.</td></tr>
<tr><td colspan="3">Original Vegetation:</td></tr>
<tr><td colspan="3">III. Dimensions of erosion</td></tr>
<tr><td colspan="3">Length: 85 m</td></tr>
<tr><td colspan="3">IV. Characteristics of the contributing area</td></tr>
<tr><td colspan="3">Use and occupation of the contributing area: Crops upstream and downstream in an area of suppressed native forest (APP).</td></tr>
<tr><td colspan="3">Causes, conditioning factors and mitigating factors: The owner, Mr. Josino, said that the property was used to grow sugar cane between the 1930s and 1940s in a very rudimentary system with animal traction and the practice of burning. Around 1962, with the arrival of the tobacco companies in the region, some of the sugarcane plantations were abandoned for tobacco cultivation, until in the 1980s the property was taken over solely for tobacco production, which has continued to this day. The origin of the ravine, according to Mr. Josino, dates back to the 1970s when there was a small furrow, ploughed every year and reopened every year during heavy rainfall. According to Mr. Josino, one year after the construction of contour lines on the site, an extreme rainfall event broke the contour line, opening up a larger furrow. At that time, the area contributing to erosion was occupied by tobacco grown in a conventional system with molchões. The owner had to abandon the area due to the deepening of the erosion, and currently on the left side the area has become a field in which the owner keeps animals and tramples on the edges of the ravine, contributing to its advance. On the direct side of the erosion, the owner continues to plant tabano in a conventional system without contour lines, directing the surface runoff from the crop into the ravine. This is a steep slope with degraded soil and a sandy texture, indicating the exposure of the saprolitic horizon. In short, the characteristics of the local physical environment are indicative of the fragility of the area in terms of the development of linear erosion in the form of furrows, which, together with the historically degrading type of land use and occupation, have triggered the emergence of the legendary gully that could evolve into a gully.</td></tr>
<tr><td colspan="3">V. Characteristics of linear erosion</td></tr>
<tr><td colspan="3">Erosion dynamics (general characteristics): The erosion, which has not yet reached the water table, is a ravine formed by a single main axis that receives runoff concentrated at various points coming from the upstream and right-hand side crops. It is more stable at the head due to the maintenance of vegetation and on the left side it is kept fallow and isolated with a fence. The slopes are steep along the entire length of the ravine on a steep slope. The presence of pioneer vegetation throughout the ravine did not make it difficult to prove erosion mechanisms such as the existence of pipings, but the evolution of erosion occurs laterally (on the left side) and steadily through the slow and continuous lowering of the slopes due to the concentrated rainfall runoff from the crops.</td></tr>
<tr><td colspan="3">Evolution forecasts: The concentrated surface runoff from the crops that extends to the edge of the ravine makes this erosion unstable and tends to evolve laterally.</td></tr>
<tr><td colspan="3">VI. Main impacts</td></tr>
<tr><td colspan="3">The outbreak of this erosion triggered the following impacts:
- Local soil loss with reduced cultivation area;
- Siltation of the stream downstream;
- Risk of accidents due to falling animals and machinery, given the close proximity to the crops.</td></tr>
<tr><td colspan="3">VII. Preventive and containment measures implemented:</td></tr>
<tr><td colspan="3">Isolation of the left side of the ravine, preventing animals from approaching.</td></tr>
<tr><td colspan="3">VIII Suggested preventive and containment measures:</td></tr>
<tr><td colspan="3">As a preventive measure against erosion, it is suggested that local land use be adapted to its carrying capacity, thus avoiding the triggering of environmental impacts of this nature, while</td></tr>
<tr><td colspan="3">that the alternatives for definitive containment aim to stabilize the erosion by isolating the area and using vegetation (native and exotic plants) until more complex engineering and bioengineering works are implemented. Among the many possibilities for intervention to stabilize erosion, the following are suggested: a) Isolation of the area within a radius equal to or greater than one meter, considering this a minimum measure for safety and to encourage spontaneous repopulation around the erosion; b) Reordering the runoff from farming by installing a greater number of properly planned and vegetated drains in places that are more resistant to erosion, as well as adopting soil management practices aimed at reducing surface runoff, such as no-till farming with contour lines and soil parcelling; c) Revegetating the headwaters of the ravine with native species typical of the region to reduce the speed of surface runoff.</td></tr>
<tr><td colspan="3">X. Identification of the card</td></tr>
<tr><td colspan="3">References Mr. Josino de Avila</td></tr>
<tr><td>Team: Juliana Dummer (master's student) Roberto Verdum (supervisor)</td><td>Location: 30°44'20.03"S 51°58'37.98"O
Altitude: 180m</td><td>Date: 20/07/2014</td></tr>
</table>

ANNEX IV: Linear erosion record sheet number 15 (map 2).

<table>
<tr><td colspan="3">EROSION REGISTRATION FORM</td></tr>
<tr><td colspan="3">I. Identification and location of erosion State: RS Municipality: Chuvisca</td></tr>
<tr><td>Description: Voçoroca bordering Rincâo</td><td colspan="2">Location: Prop. Vilson Wegner</td></tr>
<tr><td colspan="3">Access: Rincâo Road</td></tr>
<tr><td colspan="3">II. Regional data</td></tr>
<tr><td>Watershed: Arroio Sutil Watershed</td><td colspan="2">Geomorphology: Hill Compartment.</td></tr>
<tr><td>Geology: Pinheiro Machado Complex composed of whitish gray granodiorites, porphyritic gray granodiorites and veins of pink granodiorites measuring 30 cm thick.</td><td colspan="2">Pedology: (PVA2), composed of Argissolos associated with Neossolos.</td></tr>
<tr><td colspan="3">Original vegetation: typical APP forest.</td></tr>
<tr><td colspan="3">III. Dimensions of erosion</td></tr>
<tr><td colspan="3">Length: 118 m</td></tr>
<tr><td colspan="3">IV. Characteristics of the contributing area</td></tr>
<tr><td colspan="3">Use and occupation of the contributing area: Upstream, eucalyptus forest, on the sides, farming, currently in disuse with pioneer vegetation.</td></tr>
<tr><td colspan="3">Causes, conditioning factors and mitigating factors: The convex divergent shape of the slopes provides a large volume of rainfall runoff at a single concave point converging between the two slopes (talvegue) with geological weaknesses due to the exposure of the saprolitic layer with low stability and support. This runoff is accentuated by the presence of unpaved roads upstream with runoff channels directed at erosion incision</td></tr>
<tr><td colspan="3">V. Characteristics of linear erosion</td></tr>
<tr><td colspan="3">Erosion dynamics (general characteristics): erosion that has already reached the water table has three axes of development at the headland (fig. 2). The central axis is generated by the concentration of runoff from the upstream slope and the upper channel of the road in a concrete pipe. The secondary right axis advances by concentrating all the runoff from the slope at this point, as is the case with the right axis. The concentration of runoff due to the breadth of the catchment drained at a point in the terrain with a layer of exposed saprolitic soil.</td></tr>
<tr><td colspan="3">Evolution forecasts: The fact that the soil on the sides is not being turned over, and that the natural vegetation inside the gully is well developed, indicates that the risk of evolution is essentially located at the headland bordering the road. The frequent maintenance of the road provides some control of the erosion process, but the concentration of three drainage channels in the gully in extreme rainfall events can generate severe remontant erosion by increasing its length and laterally, due to future junctions of its minor axes.</td></tr>
<tr><td colspan="3">VI. Main impacts</td></tr>
<tr><td colspan="3">The outbreak of this erosion triggered the following impacts:
- Local soil loss with reduction of native forest (APP) and agricultural area (crops)
- Siltation of the stream downstream.
- Risk of accidents due to falling animals and vehicles, given the proximity of the erosion to the road, especially the school transport line.</td></tr>
<tr><td colspan="3">VII. Preventive and containment measures implemented:</td></tr>
<tr><td colspan="3">There are no preventive or containment measures in place, apart from the maintenance of the road, which is designed to stabilize the erosion process.</td></tr>
<tr><td colspan="3">VIII Suggested preventive and containment measures:</td></tr>
<tr><td colspan="3">Remove the rainwater channel from the gully and install a greater number of properly planned and vegetated gullies, avoiding an excessive volume of water at a single point in the terrain.</td></tr>
<tr><td colspan="3">X. Identification of the card</td></tr>
<tr><td colspan="3">References: Mr. Ivo Wegner</td></tr>
<tr><td>Team: Juliana Dummer (master's student); Dr. Roberto Verdum (advisor)</td><td>Location: 30°45'47.39"S 52° 0'58.26"O
Altitude: 152m</td><td>Date: 20/06/2014</td></tr>
</table>

ANNEX V: Registration form for linear erosion number 32 (map 2).

EROSION REGISTRATION FORM

I. Identification and location of erosion State: RS Municipality: Chuvisca

Description: *Linear erosion APP field* | **Location:** Mr. Ingo Peter's property

Access: *Costa do Pinheiro Road*

II. Regional data

Watershed: *Arroio Sutil Watershed* | **Geomorphology:** *Ridge Compartment*

Geology: *Pinheiro Machado Complex composed of whitish gray granodiorites, porphyritic gray granodiorites and veins of pink granodiorites measuring 30 cm thick.* | **Pedology:** *(PVA2), composed of Argissolos associated with Neossolos.*

Original vegetation: *Semideciduous seasonal forest*

III. Dimensions of erosion

Approximate length: *250 m*

IV. Characteristics of the contribution area

Use and occupation of the contributing area: *Corn fields and exposed soil, road interrupted by several erosion furrows upstream and APP (downstream of the slope).*

Causes, conditioning factors and mitigating factors: *The current resident (who has owned the property for more than 10 years) said that when he bought the property there was already erosion. According to him, the erosion has been increasing for at least 4 years, mainly due to a secondary lateral axis that runs in the direction of the property's buildings. The same erosion visited 4 years ago was more stable than it is today, indicating that throughout history this land use could have been altering the water flow and accelerating the flow towards the main channel, causing it to become vertical. In 2010, the area upstream was not fully cultivated, but is now surrounded by cornfields and areas that used to be fields are now exposed for planting. Another aggravating factor is the abandoned road for light vehicle traffic, with large numbers of erosion furrows formed by the concentrated runoff from the main road laid across it. The lack of drainage channels concentrates runoff on a slope of approximately 200 meters. This situation was confirmed by the extensive swelling of the soil within a 10-meter radius of the erosion, even 4 days after the end of the rain. As this is an area of intense agricultural use, even though part of the basin is covered with forest, in addition to agricultural use, there are two other hypotheses for the cause of erosion, since there is a lowering of the base level. One would be related to the greater amount of water entering the system, which in turn would be responsible for the acceleration of the concentrated flow process, and thus the verticalization of the drainage network. Another hypothesis could be related to the influence of neotectonics which could be slowly uplifting the Shield, activating the lineaments and accelerating the notching of the channels due to the increase in potential energy and incision by the water.*

V. Characteristics of linear erosion

Erosion dynamics (general characteristics): *The erosion has quite verticalized channels, presenting a 2 axis with a headland in the process of widening and reassembling erosion. The central axis is the most stable, while the developing secondary axis shows successive subsidence. The slopes are steep in the middle and upper reaches of the ravine.*

Evolution forecasts: *There is a lot of subsidence in the walls of the secondary axis, due to the concentrated surface runoff that provides high soil saturation around the ravine. These mechanisms give instability to the erosion pattern, which tends to evolve again and again, increasing its length and laterally, due to future junctions of its minor axes.*

VI. Main impacts

The outbreak of this erosion is currently causing the following impacts:
- *Local soil loss with reduction of field area;*
- *Siltation of the stream downstream;*
- *Risk of accidents due to falling animals (according to the owners, there have already been accidents involving animals in the area); - damage to the property's buildings due to the advance of the ravine over them.*

VII. Preventive and containment measures implemented:

There are no preventive and/or containment measures in place.

VIII Suggested preventive and containment measures:

As a preventative measure for the erosion process, it is suggested that local land use be adapted to its carrying capacity, with appropriate management using contour lines and vegetated cordons. The immediate suggestion is to isolate the site and reorganize the rainwater concentrated by the roads upstream.

X. Identification of the form

References: Mr. Ingo and Nelda Peter.

Team: *Juliana Dummer (master's student)* Location: *30°47'54.86"S* Date: 20/07/2014
Roberto Verdum (supervisor) *52° 0'18.98"O*
Altitude: 168

ANNEX VI: Registration form for linear erosion number 21 (map 2).

<table>
<tr><td colspan="3">EROSION REGISTRATION FORM</td></tr>
<tr><td>I. Identifying and locating erosionE</td><td colspan="2">State: RS Municipality: Chuvisca</td></tr>
<tr><td>Description: Voçoroca campo Taurino</td><td colspan="2">Location: Mr. Taurino's property</td></tr>
<tr><td colspan="3">Access: Costa do Sutil road near Kruschiel Market</td></tr>
<tr><td colspan="3">II. Regional data</td></tr>
<tr><td>Watershed: Arroio Sutil</td><td colspan="2">Geomorphology: Hill relief</td></tr>
<tr><td>Geology: Pinheiro Machado Complex composed of whitish gray granodiorites, porphyritic gray granodiorites and veins of pink granodiorites measuring 30 cm thick.</td><td colspan="2">Pedology: (PVA2), composed of Argissolos associated with Neossolos.</td></tr>
<tr><td colspan="3">Original vegetation: Deforested to plant crops, pasture and livestock.</td></tr>
<tr><td colspan="3">III. Dimensions of erosion</td></tr>
<tr><td colspan="3">Approximate length: 120m</td></tr>
<tr><td colspan="3">IV. Characteristics of the contribution area</td></tr>
<tr><td colspan="3">Use and occupation of the contributing area: small effluent treatment plant of a mineral water industry and unpaved public road (upstream of the erosion) and native forest (downstream).</td></tr>
<tr><td colspan="3">Causes, conditioning factors and mitigating factors: The current owner (28 years ago) said that in 1986 the site was a field and had a small ditch approximately 1m wide by 1m deep. That's when he laid some stone pavements to contain the erosion. In 1989, the owner transformed the field into a cornfield without the use of contour lines, which may have contributed to increasing the erosion process. Today the area is a field, used for livestock and for an effluent treatment plant for a mineral water industry (owned by the same owner) on the other side of the road upstream of the erosion. Indications of vegetation and the walls upstream of the gully with saturated soil suggest that the treated effluent is discharged into the field. This would explain the advance of the erosion process at the top of the slope. The owner of the area, in an effort to contain the erosion, said he had made channels near the side slopes of the erosion. This erosion has already reached the water table and is therefore a gully developed in Argissolos with a textural B horizon, which in turn is an obstacle to surface infiltration. The erosion incision is disconnected from the natural drainage network, but the road upstream and the concave convergent shape of the relief provide a high concentration of rainfall runoff. In short, the characteristics of the local physical environment are indicative of the fragility of the area in terms of the development of large-scale linear erosion, which, combined with the type of land use and occupation, led to the appearance of the gully. All the indications are that the lack of organization of the natural surface flow in rainy events and the artificial flow caused by the dumping of effluents and the trampling of animals on the edge of the gully are currently the agents accelerating the erosion process.</td></tr>
<tr><td colspan="3">V. Characteristics of linear erosion</td></tr>
<tr><td colspan="3">Erosion dynamics (general characteristics): The gully develops in a main axis that has fairly vertical walls, with lateral subsidence, but mainly at the headwaters.</td></tr>
<tr><td>Evolution forecasts: The gully is moving towards the road.</td><td colspan="2">a process of enlargement and erosion dating back to</td></tr>
<tr><td colspan="3">VI. Main impacts</td></tr>
<tr><td colspan="3">Loss of floor space;
Accidents involving people and animals.</td></tr>
<tr><td colspan="3">VII. Preventive and containment measures implemented:</td></tr>
<tr><td colspan="3">Random planting of bamboos without monitoring and maintenance.</td></tr>
<tr><td colspan="3">VIII Suggested preventive and containment measures:</td></tr>
<tr><td colspan="3">Isolation of the area;
Reorganization of rainwater runoff and effluents from the upstream treatment plant;
Slope smoothing and planting of ground cover vegetation.
Installation of retaining walls downstream.</td></tr>
<tr><td colspan="3">X. Identification of the card</td></tr>
<tr><td colspan="3">References Mr. Taurinho</td></tr>
<tr><td>Team: Juliana Dummer (master's student) Roberto Verdum (supervisor)</td><td>Location: 30°48 31.29"S 52° 4'16.37"O
Altitude: 172m</td><td>Date: 18/04/2014</td></tr>
</table>

Printed by Books on Demand GmbH, Norderstedt / Germany